1965

Hammaguir
Algeria

1966

Plesetsk
Russia

1967

San Marco
Kenya

1967

Woomera
Australia

1970

Jiuquan
China

1970

Uchinoura
Japan

1970

Kourou
French Guiana

THE ATLAS OF SPACE ROCKET LAUNCH SITES

AGI	Année Géophysique Internationale
AIT	Assembly, Integration and Testing
ASI	Agenzia Spaziale Italiana
	(Italian Space Agency)
ASLV	Augmented Satellite Launch Vehicle
BDS	Beidou Navigation Satellite System
BOR	Bespilotnyi Orbitalni Rakekoplan
	(Russian: *Unpiloted Orbital Spaceplane*)
CBERS	China Brazil Earth Resources Satellite
CNES	Centre National d'Études Spatiales
CNSA	China National Space Administration
COPUOS	Committee on the Peaceful Uses of Outer Space
CZ	Chang Zheng
	(Chinese: *Long March*, name of rocket)
DPRK	Democratic People's Republic of Korea
EEJ	Equatorial Electro Jet
ELA	Ensemble de Lancement Ariane
ELD	European Launcher Development Organisation
ELS	Ensemble de Lancement Soyuz
ELV	Ensemble de Lancement Vega
ETS	Engineering Test Satellite
FLP	First Launch Pad
GSLV	Geo Synchronous Launch Vehicle
HTP	Hydroxyl Terminated Polybutadiene
HTV	HOPE Transfer Vehicle
ICBM	Inter Continental Ballistic Missile
IGY	International Geophysical Year
ISA	Italian Space Agency
ISAS	Institute of Space and Aeronautical Science,
	later: Astronautical Science
ISRO	Indian Space Research Organisation
ISS	International Space Station
JAXA	Japan Aerospace Exploration Agency
KSLV	Korean Satellite Launch Vehicle
LCC	Launch Control Centre
MCC	Mission Control Centre
MIK	Montazhno Ispytatelny Korpus
	(Launch Assembly Building)
MLP	Mobile Launch Platform
MOL	Manned Orbiting Laboratory
MST	Mobile Service Tower
NACA	National Advisory Committee for Aeronautics
NASA	National Aeronautics and Space Administration
NASDA	National Space Development Agency
OMCF	Orbiter Maintenance and Checkout Facility
PIF	PSLV Integration Facility
PSLV	Polar Satellite Launch Vehicle
SAB	Shuttle Assembly Building
SHAR	Sriharikota High Altitude Range
SLC	Space Launch Complex
SLP	Second Launch Pad
SLS	Space Launch System
SLV	Satellite Launch Vehicle
SPROB	Solid Propellant Booster Plant
SSAB	Solid Stage Assembly Building
SSLV	Small Satellite Launch Vehicle
STEX	Static Test and Evaluation Complex
TERLS	Thumba Equatorial Rocket Launching Station
UKSA	UK Space Agency
USAF	United States Air Force
VAB	Vehicle Assembly Building
VAFM	Vandenberg Air Force Base
VLS	Veículo Lançador de Satélites
VSSC	Vikram Sarabhai Space Centre
WRE	Weapons Research Establishment
WTR	Western Test Range

THE ATLAS OF SPACE ROCKET LAUNCH SITES

Brian Harvey and Gurbir Singh *authors*
Paul Meuser *editor*
Katrin Soschinski *maps*

CONTENTS

Map of launch sites worldwide

Kodiak

1

Wallops

Vandenberg

Cape Canaveral

Gulf of
Mexico

Kourou

Alcantara

Peenemün

2

Hammaguir

A t l a n t i c

O c e a n

P a c i f i c

O c e a n

Equator   Equator

Plesetsk

3

Kapustin Yar

Black Sea

Caspian
Sea

Baikonur

iterranean
Sea

Semnan

Palmachim

Persian
Gulf

Red Sea

Vostochny

Jiuquan

Taiyuan

Sohae

Naro

East
Sea

Xichang

Uchinoura

Tanegashima

Pacific

Wenchang

Southeast
Asia Sea

Ocean

Sriharikota

Thumba Kulasekharapatnam

4

Equator

5

Sea Launch

San Marco

Indian

Ocean

Woomera

Great
Australian
Bight

Mahia

In a time when a quick scroll on Google Earth reveals a three-dimensional model of our planet on a computer screen, seeing Earth is still an awe-inspiring experience. Photographed by the Apollo 17 crew departing from the Moon, the image of the 'blue marble' reinforced our modern society's shift to an environmentally conscious species.

Source: NASA

1946

24 October 1946: The American A-4 was an American adaptation of the German war rocket, the V-2. Taking off from an American desert, it climbed just over 100 km to take Earth's first portrait.

Source: NASA

1959

14 August 1959: Our first glimpse at Earth from orbit had a rather artistic twist to it. The photographer, the US satellite Explorer 6, transmitted the historic image back to earth over a 40-minute time span.

Source: NASA

2011

Almost 40 years later, the Russian weather satellite Elektro-L No.1 captured Earth in a new resolution. Photographed through more than a simple film camera, the vibrant colours are created by the combination of visible and infrared waves of light recorded by the satellite. Today such images are mostly overlooked by the public, and it is probable you 'the reader' have never come across this image.

Source: Roscosmos

1966

The first full Earth photograph was snapped by a Soviet Molniya 1-3 communications satellite. The Molniya communications programme lasted up to the 1-93 satellite launch in 2004.

Source: NASA

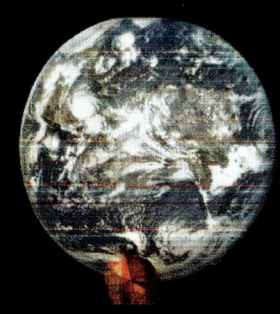

1967

On 20 September 1967, the first colour image of Earth was captured by the DODGE satellite. The satellite used a single black-and-white camera with three physical RGB filters to create the image.

Source: ISRO

2021

The Vehicle Assembly Building at the
Kennedy Space Centre on a foggy
morning in 2021.

Source: NASA, Ben Smegelsky

Unreachable space: The Nebra sky disc, found in the German town of Nebra in 1999, was designed more than 3,500 years ago. Since the beginning of cultural history, humanity has been influenced by the stars. The dream of travelling into space is within reach today.
Source: Landesmuseum für Vorgeschichte, Halle (Saale)

Paul Meuser

SPACE TRAVEL STARTS ON EARTH

The machines that orbit our planet live in a void environment – however, space travel itself does not exist in a vacuum. Travelling to space is an immense effort of humans and machines, taking not just 'one small step for man' but leaving behind a huge carbon footprint in the process. We are in the midst of a paradigm shift in which private companies and leadership figures in the form of billionaires are re-popularising space travel to an extent not seen since the space race between the USSR and USA. When thinking about space travel today, two types of images come to mind: renderings of pristine machines backlit by the blue of Earth or a large plume of smoke created by the thundering engines of a rocket leaving its launch pad. Both show in a literal sense the inherent unearthliness of space travel. Space exists isolated from the place that births its mechanical and few select human inhabitants. Thus, we tend to forget that every single thing that exits our atmosphere takes with it more than just its own weight of materials when it departs our fragile blue marble.

Beginning with the day our ancestors first began to consciously observe the starry night sky, our perspective has not always been an outlooking one. Throughout many cultures that mastered the science of astrology, they were able to use the stars as a calendar for rituals and agricultural seasons. This fascination has left its artefacts across the planet in the form of objects, carvings, and architecture. However, for these cultures, the cosmos might have always been in plain view but out of reach. Today, illuminated by our cities' blaring lights, the stars have become bleached from our nights, and we rely for the most part on digital images of distant galaxies shot on orbiting telescopes. Within the discourse of sustainability, we might not only want to return to a life connected to nature, but also begin to re-ground our perspective on both the cosmos and its new artifacts.

According to the UN's 'Scientific Assessment of Ozone Depletion' in 2018, the total effects of exhausts produced during launches account for less than 0.1 per cent of humans' effect on the atmospheric layer. However, this paints a far too narrow picture of the real consequences we are facing with an exponential increase in space travel. As a three-dimensional discipline, it is vital to acknowledge that space exploration's effects do not simply apply to Earth's enamel: its effects continue into space itself and span the globe through political, sociological, and economical factors far beyond the scope of environmental chemistry.

In the past 200 years we have seen a shift from nationalised economies to an increasing privatisation of our earthly society. This economic shift has now found itself being launched into space on million-dollar machines – into a space that has been declared public for humanity as a whole by all nations. Thus, our cosmos has to be of public concern more than ever, as it is an extension of our earthly environment that belongs to all of humanity. Today's space is a site of resource extraction in the form of digital infrastructure for the ultra-wealthy. While Earth-orbit's first inhabitant Yuri Gagarin flew to space with the sole purpose of projecting technological superiority to the Soviet Union's ideological counterpart and as a human guinea pig, proving that humans can survive such a trip, we are still in search of answers to the question of our purpose in space. When watching Jeff Bezos and Richard Branson float by their capsule windows, we are reminded that only a handful of people are behind the steering wheel of cosmic colonisation. A glimpse into the future might lie within an inconspicuous container arriving in Ukraine. Opening it reveals a wall of cardboard boxes containing the newest Starlink internet system, courtesy of SpaceX. While Europe regresses into the brutality of war, an American billionaire commands a satellite array to stream free internet to the proud new owners of the company's prototype receivers. As of June 2022, around 15,000 Starlink 'kits' have been delivered to Ukraine, offering another dimension in the modern war of information. With the rise of the private space sector, we have witnessed a new age of space exploration allowing for a rate of progress not seen before. As a new consumer market, the question of when space travel will become a part of our daily lives is no longer a question of how but a matter of when. To a great extent it already is, and it already raises fundamental questions about our own morals and efficacies. To reason about the ethics of space travel, we should not forget ourselves as inherently earthbound beings. At this very moment, SpaceX deploys its large artificial satellite constellations in Earth's orbit to create a new broadband internet system with its Starlink programme, with little regard for its environmental repercussions. The artificial mega-constellation has already proven disruptive to astronomers. While SpaceX might find it difficult to build over 2,000 radio dishes anywhere it likes here on Earth, there is little to no regulation in orbit, which continues to be run as a first-come-first-served grey zone. With space exploration as a science of escapism, it is not yet part of our perception that the anthropocene expands beyond our atmosphere. Without a clear code, it is only when treating space as an equally valuable extension of our atmospheric environment that we can question with what ease and impunity nations and companies have been able to reform it to their liking. Space has become a free for all for whoever can afford to access it. The way we conduct space flight has become a danger to us all, with space trash not just threatening space travel itself but affecting astronomy. While it might seem exciting to identify a shooting star as 'probably a satellite' on a clear night, SpaceX' Starlink satellites have already appeared on astronomical observatory images. As of now, experiencing astronomical visual imparity does not affect the general public's life, but it is the first earthly repercussion of our orbital footprints. In total more than 8,000 satellites orbit Earth, 1,807 of which were launched in 2021 – a single year accounting for an increase of around 20 per cent. Excluding the tens of thousands of pieces of space debris larger than 10 cm and countless smaller ones, the metal cloud that surrounds our planet will continue to grow while solutions or policies are yet to be implemented.

On the other hand, privatisation has also yielded a vast array of innovation, deemed almost impossible by its bureaucratic and increasingly stagnant government counterparts. When SpaceX achieved the first controlled landing of a Falcon 9 after successfully propelling its payload into orbit, the industry changed. Ever since the original mission of the space shuttle as a reusable launch platform had failed due to its immense cost, space exploration had remained an endeavour of single-use technology. With every payload consuming not just fuel but the entire launch vehicle, regular space travel was not physically, but financially impossible. With this technology now being employed by Blue Origin, and in a different form in Virgin Galactic's air launch system, they have been able to bring the price down to viably make space a permanent frontier. Although it has been leaning heavily on the private sector, which brought the price of launching a pound of payload into space from tens of thousands to a couple of thousand US dollars in just 10 years, NASA is still on track with its vow to bring the cost to hundreds of dollars within the next 30 years. Today, in the modern age of Silicon Valley and world-changing digital technologies à la Facebook, big government cannot compete with the sheer innovative output of the private sector. And it is still struggling to regulate the last two decades of innovations to our daily lives. It is this incapacity of an increasingly antiquated leadership to adapt to new industries that makes it so dangerous. This opens the door for a new generation of daring but also morally agnostic business leaders to exploit the regulatory grey zone.

An anecdote: When Palmer Luckey, the founder of the most popular VR system, Oculus Rift, sold his startup to Facebook for

Why fly into space?
What good does it do?

$2 billion in 2014, he found himself out of work with a lot of money after he left the company only three years later following a dispute over a donation to a pro-Trump group. His newest endeavour? A defence startup called Anduril Industries. While most people shy away from the ethical dilemmas of producing machines of war, the young entrepreneur saw a market gap. In the fiscal year of 2020, the US Department of Defence increased contract spending to $445.5 billion – a budget usually reserved for the industry's historic big players. And he is not alone. Being more nimble to churn out innovative and disruptive technologies, many smaller companies have figured out a way to apply Silicon Valley's lean startup approach to take their cut out of the budget. Although a private sector has always been interwoven with governmental reform, we are living in an age where technology has a never-before-seen power over our lives. With this power now being introduced to a field of weaponisation, we are now in a world where technology is void of morality. With the world around us becoming more and more dependent on private services over government, we are forgetting that we are putting our fates in the hands of profit-oriented entities and giving up the democratic rights given to most of us by the system we live in – a fact we should consider when looking at the space industry, which was born from the need for ever-more destructive intercontinental missiles with a side of science. With entities acting unregulated and in a strictly profit-oriented way, we will inevitably see the return of the patterns of colonisation and exploitation that we have witnessed repeatedly throughout human history.

What if UFOs are just billionaires from other planets?

Vanguard 1, the fourth artificial object in space, might not carry the significance or prestigious title of being the first of a kind, but it has taken on a much different role. Since its launch in 1958, it has become the oldest human-made object in space. It is an impressive feat, yet it is not its age of 63 that is of significance, but rather the precedent it has set. The satellite's last contact was in 1964, which means that it has spent the last 57 years doing nothing more than acting as a symbolic time stamp for the history of space travel. Space does not just consume billions of dollars, but it is a timeless place where the fingerprints we leave as a by-product of our presence will last for millions of years. With a science leaving such everlasting traces and being resource intensive and exclusive, we must ask ourselves: What good does it do for the unelected and unregulated few to reshape the entire orbit of our planet? What has space travel changed for humans since our first ambassador, Yuri Gagarin, was flung into orbit 60 years ago on 12 April 1961? In answering this question, we should consider that we should not condone science for the sake of science, as humanity's thirst for exploring the unknown is what has levitated our species to where it is today. Equally, we should never stop questioning the very actions we take and weigh the costs of this passion of our species.

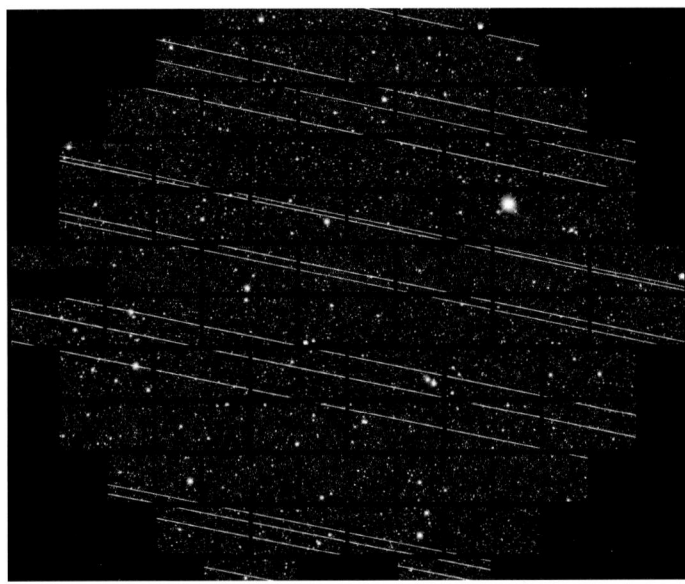

A long exposure image taken by the Cerro Tololo Inter-American Observatory (CTIO). The streaks on the image are not due to technical issues, but a series of SpaceX Starlink satellites buzzing through the night sky in a close formation.
Source: CTIO / NOIRLab / NSF / AURA / DECam DELVE Survey

Only a month after Gagarin's 108-minute venture into our orbit, President Kennedy spoke to his fellow Americans, giving his justification for stepping not just across borders, but off Earth entirely.

**Flying into Space:
The Challenge of Few or the Nature of Mankind?**

'We choose to go to the moon in this decade and do the other things, not because they are easy, but because they are hard, because that goal will serve to organise and measure the best of our energies and skills, because that challenge is one that we are willing to accept, one we are unwilling to postpone, and one which we intend to win, and the others, too.' Today this JFK quote still resonates with the forerunners of space exploration. It is a strong quote that acts as a rallying call more than half a decade after being spoken aloud. However, we should refrain from accepting such fallacious reasoning, even if it is given under the pretence of a unified goal to conquer space as a united species. There should be no doubt: We went to the Moon because two nuclear superpowers were aiming to prove their military superiority. This facade of space travel as a science for all of Earth's inhabitants has allowed it to avoid the scrutiny that political and economic endeavours are usually destined to be exposed to. We have to break from thinking purely scientifically and militaristically and open the conversation to economical, sociological, and ethical considerations – an act that has historically only come after whatever consequential actions and decisions we have made as a species.

Today the discussion about the purpose of space travel has been side-tracked by a tabloid-like news cycle of billionaires 'spending' their way into the history books. With a shift in public perception away from government-funded space programmes to its new private competitors, the accountability towards the taxpayer

(representing the nations as a whole) has shifted towards individual consumers, as a paying individual. The difference between space for all or the individual makes for an immense ideological difference in the forms of what is demanded and supplied. The measure of success in space is no longer just measured by milestones or the scientific benefit it returns, but the return on investment it can achieve. And so the question: If space contains anything with inherent value to our society, has this been forgotten entirely in the eyes of a customer-experience focused industry? Will a 'Disney cooperation' of space entertainment be fit to lead the costly endeavour into this frontier? On the other hand, as diabolical as it might be, shooting the world's most powerful economic leaders into our orbit or beyond might be just the thing we need. When astronauts look back on their time in space, they all bring back a universal epiphany: We are small, among the billions of galaxies, and our precious blue wonder that we call Earth is one enclosed organism. What better or more humbling experience can our Earth's wealthiest experience than the psychological shift in perspective they gain from this almost outer body experience. So how come neither Richard Branson nor Jeff Bezos has returned as a reborn Greta Thunberg? With a $2.54-million-a-minute price tag on one's mind aboard a Blue Origin capsule, we have yet to see the psychological impact it has on its passengers. It can be said, however, that the epiphanies inside Earth's first astro-billionaire's minds have not yet manifested themselves into change on Earth. 'Seriously, for every Amazon customer out there and every Amazon employee, thank you from the bottom of my heart very much. It's very appreciated.' Those were the words Earth's inhabitants received after the Amazon founder's personal $5.5bn space endeavour. Words that struck a chord with the people affected by the gruelling working conditions that remain a point of contention at Amazon's facilities. Countering the criticism, Bezos pledged to spend $10bn through his 'Bezos Earth Fund' by 2030. We should always demand an impact and earthly change in the right direction without financial exclusion. And so, we need to demand more than just space travel as the world's most expensive life-changing rollercoaster ride. What good does it do to keep the exclusive power of space accessibility in the hands of the elite and the few individuals they grant their generosity to? The answer is not to shoot into space whoever sits atop the exclusive waiting list, but rather to begin to not take for granted the hidden infrastructure that lies outside the windows of tourist capsules.

The golden ticket: it is no wonder that it is in exactly this sector that the entertainment industry has found its newest home. No other science evokes such awe in children and adults alike, and how could they in the face of such mystic images of explosions, rockets, and space objects. It is a science disconnected from everyday society but as accessible as no other. We cannot put ourselves in the minds of a quantum physicist sitting at the control panel of the CERN particle accelerator, but we can sit in the cockpit of a space shuttle – how hard could it be?

A View of Space: Understanding our Place?

The danger in the perspective of the shuttle commander lies within the view we decide to embrace. We are looking outwards towards the stars, ready to be shot into the unknown with no

A picture posted via Twitter by the Vice Prime Minister of Ukraine, who is currently doubling as the Minister of Digital Transformation. Shown on it is a truck of Starlink Satelite Antennas promising free internet to its users in a new digital information war.
Source: Mykhailo Fedorov / @FedorovMykhailo / Twitter

time to look back. We look past what it took for humans to step onto the lunar surface in 1969. When we reason about space travel, we should refrain from focusing on the human footprint in grey dust and instead look at Apollo 8's Earthrise and Apollo 17's Blue Marble. Both are snapshots that were not meant to be, as it was not part of the mission protocol to waste film on documenting our home planet, rather anything but that. Images of Earth were part of the vaguely labelled 'targets of opportunity' category and as such took the lowest priority. 'We didn't think about that (Earth),' was the concerning recollection of William Anders, one of Apollo 8's astronauts. But despite all the odds, the images turned into the two most influential pictures ever taken from a human as well as environmental perspective.

And just like that, two pictures that almost had not been taken became symbols of a new consciousness of Earth and its resources. Triggered by the aftermath of two world wars, impending nuclear annihilation, and an alarming increase in catastrophic natural disasters caused by humans, the contemporary environmental movement was born. Catering to these newly formed counterculture communities, the Whole Earth Catalog was created by Stewart Brandwhich. It became a manifesto for rethinking our way of life through products, technology, and theories. The cover featured a portrait of Earth. Its idea was to provide the new ideological settlers with modern tools to use to sustain their new-found autonomous lifestyles. The cover was a deliberate choice. It was the power of the photographic images of Earth that solely transcended the Cold War space frenzy and brought back to Earth something bigger than national bragging rights. They profoundly and irreversibly changed the way humans think about the environment.

Earthly solutions require an inward perspective. This stands in direct contradiction to NASA's search for earthly (in their own words 'American') solutions in 'air and space'. A mission statement that, by its own definition, fosters a way of thought that focuses only on contributions it returns to Earth, in disregard for what it takes to space. During Yuri Gagarin's flight, he is said to have stated: 'I looked and looked and looked, but I didn't see God.' It is unclear if these were his words or merely a propagandistic creation of the USSR. However, it takes an immense act of human (godlike) intervention to sustain not just technology but life in the radiation-ridden vacuum of space. Space as a new frontier of the twentieth and twenty-first centuries presents us not with new moral quarrels, but with the same overlooked ethical decisions we have faced in our lengthy history of colonisation and environmental destruction. We cannot let our ambitions be governed by only the limitations of the billion-dollar budgets assigned to this expedition. It might suit them well to look back on the 50-year-old Whole Earth Catalog that called for its readers to recognise their status as 'gods' in the context of saving Earth rather than polluting and abandoning it. Space is a blank canvas on which we are already inscribing our irreversible marks based on our existing habits.

Money Making:
Private Financing of State Responsibilities

Progress always comes at a cost. However, the process of lifting 12 tonnes of rocket 400 km into space to supply our human outpost in the form of the International Space Station does not start with the struggles of astronauts in space, but rather the supply networks and political systems that roll the rocket onto the launch pad in the first place. When addressing issues of logistics for space travel, the unit of measurement is defined by the payload capacity of the utmost tip of a rocket. In 2018, Elon Musk placed his

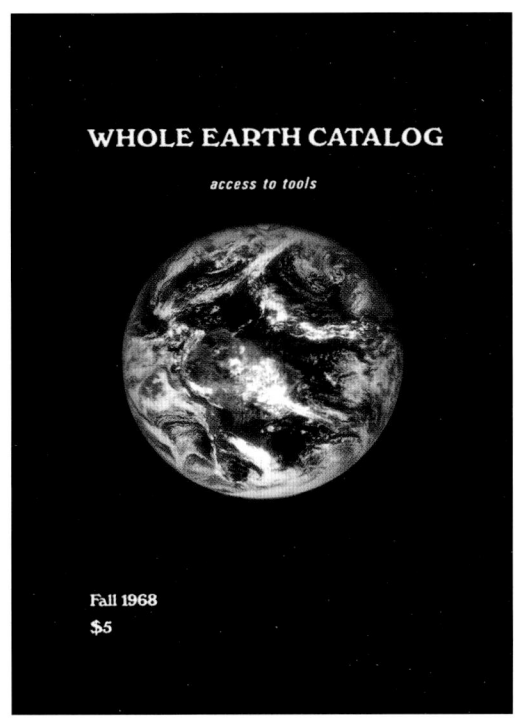

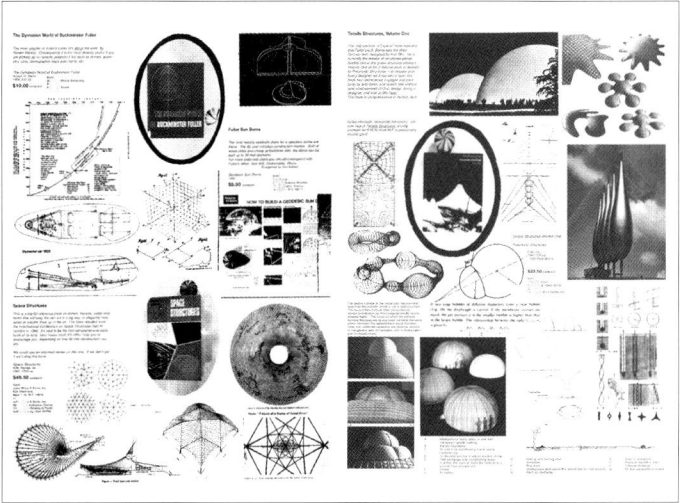

The Whole Earth Catalog combined ideological texts and a new form of product review tailoring to the modern environmental movement and an alternative, independent lifestyle. Pre-internet, it taught its readers how to use both old and new technologies for a sustainable way of living. Source: Whole Earth Catalog / monoskop.org

red Tesla cabriolet inside SpaceX's Falcon heavy rocket as a dummy payload. As a car it represents not just an arbitrary product of human industrial capability, but one of its largest. In many ways the car industry is responsible for a large portion of today's carbon emissions. It is an industry that has only recently acted on its involuntary responsibilities towards sustaining life on Earth. In the next 20 years, we will see immense growth in electric cars, shifting the environmental strain from carbon emissions and fossil fuels to new complex supply chains and hazardous battery waste. The red electric car in this scenario is strapped to a rocket consuming around 1,500 kg of fuel per second, emitting carbon emissions equivalent to around 200 years' worth of what a regular gasoline-powered car would emit. We can hardly call an electric car strapped to a launch vehicle a green endeavour.

Any satellite taking the place of the red Tesla is a product of the same supply chain. A system of mines extracting ores, oil rigs pumping oil, and resources being sent to refineries to finally complete the international web of resource flows into a single product. A rocket payload, just like a cabriolet dealership, masks the gory and messy means of production hidden behind its polished surfaces. We cannot see our presence in space with our bare eyes; it is a place shown only through carefully curated images of unfathomable machines. To begin to answer the new socio-economic questions raised by our rapid expansion into the void, we need to look no further than follow the cracks in the concrete of our planetary launch sites. The rusty train tracks leading to the pad break the pristine and sterile looks of space and reopen our eyes to the fact that even this engineered endeavour has its flaws.

The Cassini spacecraft began its journey in 1997 from Cape Canaveral Air Force Station and covered the 1,200m km between Earth and Saturn in around seven years. It stayed in orbit until it

was intentionally crashed at its mission's end in 2017. It undoubtedly returned invaluable scientific data, including more than half a million pictures of Saturn and its moons. Space exploration's carbon cycle works very differently to the one we have become accustomed to. The 5,700 kg of satellite that burned up in Saturn's atmosphere were part of a complete 860,000 kg launch assembly. A large amount of fuel and material for a one-off science endeavour, but a fact that begs the question of scalability when 1.4m kg Falcon heavy rockets begin flying 70 missions a year by 2023. This book begins to give a glimpse into the infrastructure constructed internationally that is needed to sustain our ongoing presence in space. While the vast majority of launches are dedicated to sustaining our orbital infrastructure of satellites, the ISS receives a resupply mission nine times a year. The current human population in space consists of six ISS astronauts and three astronauts aboard the Chinese Tiangong space station. Supplying an average-sized family (with 4.9 being the global average) with a full-size rocket every 45 days will become an immense logistical struggle, especially considering that the ISS orbits Earth at an altitude of 408 km, while the distances between Earth and the Moon and Mars are 384,000 km and 372m km, respectively. While the idea of reusable rockets has been a concept used as far back as the space shuttle, we should expect a disparity between the resources we send versus what is returned.

As with other 'space dilemmas', we do not have to look far into our earthly past to find an appropriate case study. The world's (unofficial) fifth largest economy as of 2021 is the US state of California. This, however, was not always the case, as 'Alta California', as it was first known under the Spanish, was a harsh and wild place, cut off from its colonial powers by both the Rocky Mountains and the American continent (pre-Panama canal). Under the eyes

Space has become an industry. An image tweeted in 2021 shows SpaceX engineers underneath the new Starship being paired with their Super Heavy rocket. Source: Elon Musk / @elonmusk / Twitter

of the native population, control over the US Pacific sea border was fought through a mainly economic war between Russia and Spain, which was later succeeded by independent Mexico. The foothold provided by missionaries and strategic forts caused great economic burden for the Spanish and Mexican states, with outposts being dependent on the regular supplies sent from Mexico. Just like our future space endeavours, these outposts were held alive by brute force and will to be finally annexed into the mainland. Mexico, which had just freed itself from its own colonialist supervision, was keen to expand its influence to the north. This made it more infuriating when after three centuries of unrewarded colonisation, the land was ceded to the United States following the Mexican-American War. To add insult to injury, the loss coincided with the discovery of gold in the new US territory, incentivising a frenzy of gold-hungry miners to make the hazardous trek to the far west. This triggered an explosive economic boom that had failed to materialise in the past 300 years of Mexican rule. As the moment of the first human on Mars gets closer, we might be preoccupied with the logistics of the first few settlers – too occupied to see the vultures lurking in the shadows of achievement. Anticipating a long struggle for a foothold on our red solar neighbour, a modern gold rush for resources will be triggered if the valuable resources become accessible.

How will we flex our muscles in space once accessibility is no longer a hurdle? When the United States launched its newest military branch, it was laughed at and belittled. But there is a difference between fighting a secretive war of surveillance satellites, space manoeuvre games of chicken, and eyeing our opponents through the radar screen to calling to arms a military branch to 'protect, and expand the US fleet' in space. Modern warfare has entered new altitudes, and that is not new news; flying to the Moon atop the Saturn V rocket (the rocket used in the Apollo missions) was once considered of equal importance as carrying nuclear ordnances. In 2007, China launched a missile into space, shattering their own satellite and creating a debris cloud that is still trackable today. One year before the US Space Force was announced, NATO added space to its contemporary operational domains of air, land, sea, and cyberspace. And as of 2021, Germany has stepped up to take on the task of defending space with the creation of the Weltraumkommando der Bundeswehr or WRKdoBw for short – a name less snappy than its American counterpart, but surely as efficient as an account password. Defence by deterioration has and will always be a battle of nations who can afford it. With the increasing role of private space corporations, we are also at risk of returning to capitalism's early endeavours into private colonisation.

Colonisation has an impact on the people it suppresses. Looking past that due to the inherent void of a live environment on both the Moon and Mars leads to many overlooking the past history's precedents. The disparity created by the exclusivity of interstellar colonisation will be brought back to our planet in the form of asymmetrical distribution of wealth and resources. Today we still experience the effects of history's mercenary and colonial footprints, manifesting themselves in the chain of port cities that still harbour the globalised trade of today. However, this remains a flawed measure of the effects of interstellar commerce. Despite our supply chains being stretched across oceans and continents, their centres of distribution are still grounded on our planet's surface. While we should be focusing on limiting or eliminating the carbon emissions of our globalised trade, we are about to embark on introducing an entirely new dimensionality and infrastructure to the equation. With the viability of rockets as rapid means of commercial transportation, we are about to enter a world where space travel and earthly infrastructure become more intertwined than ever. In 2020, 114 rockets were launched to space, with over 70 per cent being of either Chinese or American

Unlike the iconic 'Lunch Atop A Skyscraper' photo, Elon Musk's tweet might not have had the same cultural impact, but the message is clear. While the workers atop the beam of the Rockefeller Centre represented the country's resilience during the Great Depression, the skyscraper became a symbol of technological achievement. Today the rocket is the technology behind which nations can rally.
Source: Library of Congress

origin. Looking at these numbers, we quickly realise that space is not the shared global utopia it is advertised as. It is as such not too far-fetched to expect that in order to sustain an ever-growing space presence in combination with a potential global rocket network, we will increase the launch pad traffic to potentially hundreds or thousands of launches a day. While countries fight for every inch of coastal territory or polar drilling rights through legal and militaristic shows of force, space remains a numbers game reserved for the power that can claim as much physical space as possible for its presence. With the only physical trace of this conflict being tiny specs in our night sky filtered to society through illustrious and exuberant articles, it can be easy to overlook the dangerous long-term trends we are setting.

About this Atlas:
Mapping the World's Space Rocket Launch Sites

This book offers a unique look at the physical footprints of Earth's launch sites. With most places hidden away in jungles, deserts, or amid the Central Asian steppes, these places exist for the most part out of the eye of the general public. With satellites facilitating our modern society and a modern space age ever-present in today's news cycle, it is now more important than ever to think about the imprint these undertakings leave on Earth. The future of space exploration is unfolding in front of our eyes one launch at a time. Consequently, this Atlas of Space Rocket Launch Sites should provide the reader with an in-depth historical understanding and offer a glimpse at the geographic and geopolitical lines these launch sites occupy. On the following pages, the order of the launch sites follows the Earth's rotation west to east, from the Alaskan north-west to a floating launch pad in the eastern Pacific. This reading results in five geographical regions sorted

regardless of their political affiliation and reveals a surprisingly eastern-centric accumulation of launch sites. In sum, the Americas chapter introduces not just the world's most famous launch site, but expands from the US Pacific coast to the Atlantic jungles of Brazil. This is followed by the Balto-Mediterranean triangle and the landmass of Eurasia. These sites witnessed both the dark beginnings of the first ever rocket launch by Nazi Germany on the coast of the Baltic Sea (Peenemünde) and the ground-breaking history of humanity's first satellite, Sputnik, starting in the steppes of Soviet Kazakhstan (Baikonur), among many others. The Indo-African region, bordering the Indian Ocean, is lesser-known, with launch sites in Kenya and India. The last chapter is dedicated to the Asia-Pacific region, which hosts the most space rocket launch sites by number.

Each place has been researched in detail. Some information was easy to access in archives and earlier publications, while some sites have been forgotten or are still shrouded in the secrecy of militarised space programmes. All the maps have been precisely drawn in a scaled manner based on the collected materials in order to be read in an understandable way. For the first time ever, a comprehensive collection of mapped launch sites is presented in this atlas. The additional illustrations aim to help the reader understand the places on this planet that they may never be able to visit. My thanks go to Brian Harvey, whose historical knowledge and rigorous research form the foundation of this book. It has been a great journey from our cooperation on the *Architectural Guide: Moon* (DOM publishers, 2019) back to the beginnings of our interstellar ambitions here on Earth. Gurbir Singh contributed to this volume with his expertise about the launch sites in India. And finally, this book would not be the same without Katrin Soschinski and the energy and professionalism she put into the maps.

Robert Hutchings Goddard with his first liquid-
fuelled rockets in Auburn, Massachusetts.
Source: NASA

Brian Harvey

AN INTRODUCTION TO THE HISTORY OF LAUNCH SITES

The first rocket pioneers did not have the gleaming high-tech launch bases we know today. Sergei Korolev and his colleagues headed out into the forests around Moscow and let off their small rockets in clearings, hoping that they would at least clear the tree-tops. Robert Goddard fired his small rockets from a farm, now a golf course, in Auburn, Massachusetts. Wernher von Braun and his colleagues made a *Raketenflugplatz* ('rocket flight place') in Berlin itself until he was taken on by the German Army and the test grounds moved out of Berlin to the gently rolling fields and woods near Kummersdorf. Hideo Itokawa used a beach in Japan and Vikram Sarabhai operated from a beachside church – with the congregation's permission. Nowadays, a plaque may mark these places but there is not much evidence now of what took place nearly one hundred years ago. They fired quite small rockets – Itokawa's were so small they were called 'pencils'.

The first real, modern rocket was Von Braun's A-4 or V-2, which weighed over 20 tonnes, used a high-performance engine, and reached the edge of space. It required a concrete launch platform called a *Prüfstand* ('test stand'), fuelling facilities, and systems for supervision and tracking – what we would now call mission control. Homes were constructed to house the launch site workers. Thus was created the first modern launch base: Peenemünde on the Baltic Sea. Peenemünde had many of the defining characteristics of most of the world's subsequent launch sites. Being on the Baltic, the land was flat, making rail and road transport easier. It was beside the sea, so that rockets would quickly be over the sea where, if things went wrong, they would crash there rather than fall on people's homes. From the military point of view, it was distant from prying photo reconnaissance planes, though in practice distance never stopped inquisitive aircraft from discovering launch sites, as other countries were to find out later. Peenemünde was the first, but far from the last, of the 'secret' launch sites and many came to be surrounded in mystery and intrigue, their locations concealed, and their existence even denied. Seaside or coastal launch sites became the norm for the United States (Cape Canaveral, Vandenberg, Wallops Island), Japan (Tanegashima, Uchinoura Kagoshima), India (Sriharikota), Europe (Kourou), and China (Wenchang).

The A-4 was the basis of post-war rocketry, with the Allies descending on Germany to scour the country for its remains. They were able to scavenge enough A-4s and parts to launch their own. With a limited run of firings, the immediate post-war sites were minimalist, comprising a concrete pad and some tracking systems. The British fired A-4s over the North Sea from Cuxhaven; the Americans went to the deserts of White Sands, New Mexico; the British moved to Woomera, Australia; and the Soviet Union went to the flatlands east of the Volga River at Kapustin Yar. This introduced the second type of launch site: the desert. Although there was still the danger that rockets could go off course toward a populated area – one A-4 famously crashed near El Paso – most rockets or their stages were likely to fall over relatively unpopulated areas, though not always with much regard for the small numbers of people who did live there. Deserts were more challenging for communications, but being far inland, had the advantage of being even further away from observation by other countries. The desert launch base was to be the norm for the Soviet Union (Kapustin Yar), France (Hammaguir, Algeria), China (Jiuquan), Iran (Semnan), and Britain (Woomera, Australia).

By the 1950s, the world's rocketeers had largely exhausted the possibilities of the A-4 and moved to the next quantum step forward: the Intercontinental Ballistic Missile (ICBM), with significantly more powerful engines, intended for warheads, but also capable of putting a satellite into Earth orbit. The United States developed such a range of ICBMs as to require a multiplicity of launch pads, so many that they were called 'missile row'. Here, Cape Canaveral, a sand spit running along the Florida coast, already had an air force station, so land to the north was cleared for missiles. In the Soviet Union, Kapustin Yar was simply so close to American and British reconnaissance planes that a new launch site was built far in the Kazakh desert. Its location was not disclosed until the 1960s, although American spy planes found it long before, its eavesdropping stations listening in to its launch radio traffic and they realised that its official name – Baikonur – was a deception, for it was nowhere near the railhead of that name. Both Cape Canaveral and Baikonur required an infrastructure far more substantial than Peenemünde-type locations. They needed preparation, assembly, and integration buildings; clean rooms; fuel storage facilities; launch control centres; search and recovery forces; and defence systems to guard them. The actual launch pads became challenging feats of engineering, designed to enshroud and protect the rocket as it was prepared for launch, while enabling launch crews to inspect and check the rocket. Cape Canaveral's missile row became a line of red-rust-painted gantry structures stretching to the horizon. In the case of Baikonur, the Soviet ICBM had a launch system that was both complicated and simple: complicated because it required an elaborate structure of four metallic arms called the tulipan ('tulip') because of its shape; a multi-storey tower and lift; and an underground servicing cabin; simple because when thrust exceeded weight, the arms simply fell back to release the rocket. It also required the excavation of a huge flame trench and the building of the pillars for the concrete support structure – a visible godsend to spying aircraft that could find it without difficulty.

Launch sites entered the public consciousness at around this time, Cape Canaveral being the model. Like the Russians, the Americans conducted their first satellite launch attempts in secret, or at least attempted to do so. Public and press interest quickly overwhelmed the uneven struggle, so the Americans made a virtue of necessity and President Eisenhower and Congress put the space programme under a civilian agency, the National Aeronautics and Space Administration (NASA), in 1958. That year, the Americans began a frantic – and unsuccessful – series of launches to try to beat the Russians to sending robotic spacecraft to the Moon: the Pioneer series. This was the world's popular introduction to the idea of the launch site. The press – print, radio, and now television – encamped in the new motels on Cocoa Beach around the launch pads and breathlessly broadcast the Pioneer launches. Within a short time, everyone knew about Cape Canaveral. It was no longer an obscure coastal sand spit. They got used to the new lexicon of mission control, launch pads, countdowns, gantries, and blast offs. Pioneer was not successful, but there was no turning back now. *National Geographic* magazine, one of the most widely read periodicals in the United States, captured the romance of the launch site and the spirit of the time in a famous feature entitled 'Cape Canaveral's thousand mile shooting gallery'.

1963: Astronaut Gordon Cooper's Mercury-Atlas 9 rocket on one of Cape Canaveral's missile row pads, set to carry him to be the fourth American in orbit. By 1964 ,the rocket type beneath his black capsule would be stationed at around 132 strategic Atlas missile sites across the US, armed with nuclear warheads.
Source: NASA

An R7 rocket on its way to a combat position. Unlike its successors, this rocket required hours to be prepped for launch and was only deployed at six sites, setting a deadly precedent for what was to come in the Cold War.
Source: Polytech Moscow

Mercury Atlas D

SM-65 Atlas

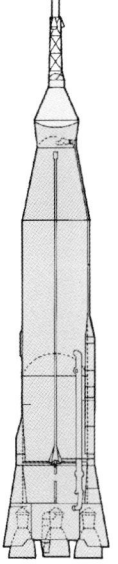
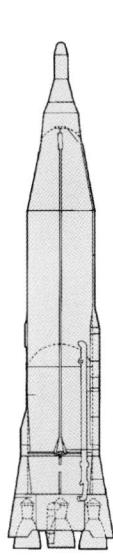

The Mercury programme used the Atlas D to launch America's first astronauts into space. The rocket was also modified to function as an intercontinental ballistic missile.
Source: NASA

R-7 Sputnik

R-7 ICBM

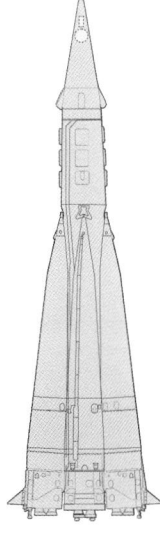
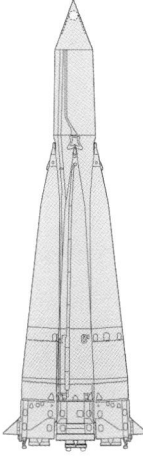

4 October 1957: The R-7 made history by carrying the world's first man-made satellite into orbit. However, its second and often overlooked accomplishment was becoming the world's first intercontinental ballistic missile. This is a modified R-7 containing the Sputnik satellite.
Source: NASA

This was only the start. In 1961, the United States began human-piloted space launches with a souped-up A-4 called the Redstone (pad 5) and orbital flights the following year with the Atlas ICBM (pad 14). NASA struggled to keep up with the level of media interest, providing press centres, viewing areas, and facilities for radio and television. Not only that, but ordinary folks could come and watch the spectacle from the beaches opposite the launch site at a safe distance. Lengthy and often frustrating launch delays meant that there was more than adequate time for the media to explain the geography of the rapidly-expanding Cape Canaveral. Launches acquired a well-known geographic choreography: the crew quarters where astronauts had their pre-flight steak-and-eggs breakfast; the walk-out to the bus in the silvery spacesuit and air-conditioning box; the bus ride to the pad; entering the cabin many floors above ground; the gantry rolling back; mission control counting off the seconds; the blast of light of lift-off and the rocket rising above the smoke as viewers shielded their eyes from the sun as the rocket curved over in its climb to head down over the blue Atlantic.

The Apollo programme meant an enormous expansion of Cape Canaveral. It required the clearing of a similar area of land to the north, beyond the lagoons. The skyline changed. Mission control – a small blockhouse – was gone, replaced by much bigger facilities thousands of miles away in Houston, Texas. To the north rose Launch Complex 39, a set of double pads erected on cement mounds; two gigantic paths for crawlers to crawl with their giant Saturn V rockets; the utterly enormous Vehicle Assembly Building, able to house four Saturn Vs at a time and one of the most recognisable buildings on our planet; the glassy launch control centre; and viewing areas. The Moon landing was the most televised event in history, from the launch from Cape Canaveral on 16 July 1969 to the landing on the Moon four days later and the return of the astronauts to Earth. No less than a million people travelled by car, bus, and camper to watch the departure from our planet. Cape Canaveral was the place where the history of that great journey started out. The launch that Cape Canaveral visitors remember best was Apollo 17, the last and the only one at night. The Saturn V lit up the night sky, with the buildings along the cape casting long shadows. As it ascended, the sound of the rocket made the ground tremble and buildings shake while the Saturn's long orange-yellow flame could be seen far along the east coast and then came the pure blue flame of the hydrogen of its second stage.

Those buildings still stand today, though they were altered in the early 1970s to accommodate the space shuttle. This required the construction, further north, of a long runway to which the shuttle would return on leaving orbit. Sadly, Cape Canaveral came to know pain when the *Challenger* exploded there on 28 January 1986. Years later, viewers waited for the return of *Columbia* on 1 February 2003 and when it did not return exactly on time, as it always did, and as the seconds ticked past, they knew it would now not be returning at all. It is right that Cape Canaveral has a fitting memorial to the brave astronauts who were lost.

As the shuttles retired and along with them the glamour of human-crewed launches, Cape Canaveral continued to serve as America's principal launch site for the routine but important robotic launches, not least of which were the series of robotic probes that explored

1967: Apollo Saturn IB stages at the Michoud Assembly Facility in New Orleans. All first stages of Saturn rockets were either manufactured here or in Huntsville, Alabama. The rockets were then transported via an ocean-going barge to their launch site in Florida.
Source: NASA

The US military-industrial complex in full swing, showing off an Atlas missile assembly line at the aircraft manufacturing company Convair in San Diego.
Source: NASA

the depths of the solar system and beyond. Cape Canaveral retained its place in contemporary American consciousness as the working launch site, the place where history was made, with popular interest and visitors unabated.

Its great rival, Baikonur, emerged from the shadows only gradually. French President General Charles de Gaulle was a privileged visitor there in 1966, so its existence could no longer be denied. Baikonur experienced an expansion in the 1960s similar to Cape Canaveral as the Soviet Union sought to stay in the race to the Moon. First came launch sites for the Proton rocket for round-the-Moon missions and then a site for the huge but unsuccessful N-1 rocket. The shuttle development at Cape Canaveral was imitated at Baikonur with the conversion of the N-1 pads for the Buran space shuttle and its adjacent long runway.

Baikonur opened up with ever more Western visitors from the 1970s, especially with the arrival of the Americans in advance of the Apollo-Soyuz Test Project in 1975. One of the benefits of their participation to the Americans was that they got a grandstand, close-up view of the Soviet space programme and its landing site in Arkalyk, also in Kazakhstan. For the first time, the launch and landing were televised, so a wider audience got a picture of how Baikonur worked, even more so when the French arrived there in 1982 for the first joint Soviet-French mission. Between 1977 and 1991, nationals from 18 other countries flew to space as part of the series that included Rakesh Sharma from India, Phạm Tuân from Vietnam, and Helen Sharman from Britain. The huge scale and distances between its facilities became evident, making Cape Canaveral look compact by comparison.

The end of the Soviet period saw *glasnost* ('openness'), which gradually made the programme and its history ever more accessible. The other big cosmodrome, Plesetsk, up in the Arctic near Murmansk, was still off-limits. It was – and remains – the world's busiest launch site, with more rockets launching from there than any other location. Plesetsk was a Cold War creation and the Soviet Union's original ICBM missile base where its nuclear deterrent was stationed, ready to fly across the North Pole and descend on North America. From the beginning of the 1960s decade, the Americans believed that the Soviet Union would station a base there and launched U-2 spy planes from Turkey to follow the train tracks to find it. While looking for Plesetsk, U-2 pilot Gary Powers reached Sverdlovsk in the Urals where he was shot down.

The Americans found Plesetsk soon enough, but worked hard to make sure that the Soviet Union did not know what they knew. The problem arose when the Soviet Union started making military satellite launches from there. Space enthusiasts in Britain, notably Phil Clark, were able to use published information on its orbit to back-calculate the exact location of its launch site, to the consternation of Western intelligence services who interrogated him on his unanticipatedly obtained knowledge. Plesetsk quickly came to be used for civilian launches, even of Western satellites, but visitors were still not permitted. Satellites must be handed over at the border and that was the last they saw of them until they were told that they were in orbit, an approach which applied to the USSR's own scientists as well. Sometimes they were told launches were from Kapustin Yar (which officially did exist), even when they were launched from Plesetsk (which officially did not exist).

The space tourism of the early days was for a more grounded type of crowd. Spectators camped on beaches and roads around Kennedy Space Centre to get a glimpse of the historic Apollo 11 launch.
Source: NASA

The secrecy in which the cosmodromes had been enveloped eventually collapsed under the combined weight of *glasnost* and the commercialisation of the space programme in the 1990s. The Russian space programme now entered its most difficult and challenging phase: funding collapsed, over half the workforce left, maintenance stopped, and the programme almost succumbed. Conditions in the cosmodromes became dire. Electricity was cut off when bills were not paid and one probe to Mars was completed by candlelight. The building housing the Buran space shuttle actually collapsed. Parts of the town of Baikonur were deserted when the inhabitants left and weeds grew. Rubbish and debris were uncollected, and everything took on a shoddy, neglected appearance. The world could see all of this, for European and American holiday companies even got permission to fly tourists to Baikonur, which was demilitarised, with soldiers and guards withdrawn.

The Russian space programme and its cosmodromes now depended on the ability to fly, for a price, Western satellite cargoes and later astronauts. These companies then built, at their own expense, new handling facilities at Baikonur, so the physical infrastructure at last began to improve. European companies insisted on accompanying their precious cargoes to the assembly halls and launch sites. With Bion 10 (1995), European scientists finally got into Plesetsk, and with Foton 7 (1997), they got to the landing site. Starting with Western flights from Baikonur to Mir and then the International Space Station, families were permitted to accompany their loved ones in the final stages of preparation before launch. They could walk along the Soyuz rocket as it rumbled along the railway line to the pad, attend the final press conference (behind

glass to avoid infections), see the astronauts walk out to the bus to bring them to the pad, and then watch them board their rocket as the Soyuz hissed liquid oxygen vapour behind them. American astronaut Leroy Chiao, who flew from both locations, contrasted the sterile environment of Cape Canaveral with the joyous party atmosphere that accompanied launches from Baikonur.

There could be no greater contrast in conditions between the situation in Russia and the great newcomer to the space world: China. Its original launch base was in the north-west desert, near Jiuquan, at the end of the Great Wall, intentionally distant from American spy planes, which soon found it anyway. It very much followed a rail-based Soviet design and layout, being built during the period of Sino-Soviet cooperation (1954–1960). The first Western visitors were from Sweden, accompanying their Freja satellite in 1992. In that year, China decided to begin a crewed space programme, which led to a substantial expansion in its facilities, most notably a vehicle assembly building not that different to the American one at Cape Canaveral, with a roller-way to two adjacent pads. Like at Baikonur, Chinese human launches were colourful events, with bands playing and crowds assembling to cheer on the crews as they headed down to the pad.

China's space programme expanded rapidly and Jiuquan was followed by no less than three new launch sites. For the launches of communication satellites to 24-hour equatorial orbit, China required a more southerly launch site, so a picturesque location was chosen in the mountains and valleys of Sichuan and opened in 1984, with two pads for the powerful Long March CZ-3 rockets. A small launch site was opened at the military base of Taiyuan in

What goes up must come down (and hopefully in one piece): On 14 April 1981,
the crowds at Rogers Dry Lake in the US crossed their fingers that that saying
would apply to the landing of *Columbia*, the first space shuttle.
Source: NASA

1988. Finally, in its biggest project ever, a new site was opened
in Wenchang on the eastern side of the island of Hainan in the
very south of China, east of the Gulf of Tonkin. This was the
true launch site for the future, designed to take the even more
powerful Long March CZ-7 and CZ-5 rockets essential to build
China's space station and send spaceships to the Moon and Mars.
In the late 2020s, it will be used for the Long March CZ-9 to send
China's first astronauts, or *hangtianyuan*, to the Moon. The rockets are too big for the rail system, so they have to be brought in
by cargo ships from the huge Tianjin factory near Beijing where
they are made.

Glasnost came to the Chinese space programme, too. Visitors could
pay €200 to go to Xichang for a seat in a viewing platform to
watch the launch of the Chang e Moon probe in 2007. The most
open launch site is Wenchang. Visitors can come to luxury hotels
close to Wenchang launch site and watch the Long March CZ-5s
light up the night sky and head toward the Philippines on their
path to orbit. The overwhelming impression of the Chinese space
programme is the newness of its facilities, like its cosmodromes,
and the youth of the people who work therein, typically in their
twenties. Its test facilities, design centres, administrative offices,
factories, and even its museums are modern steel-and-glass buildings speaking volumes of the country's ambitions to be the world's
leading spacefaring nation. It was not until the late 2010s that
Russia found the resources to once again build new facilities: the
new showcase cosmodrome in the far east, Vostochny, and its adjacent city of Tsiolkovsky. Between them, Wenchang and Vostochny
represent the latest word in launch centres.

China, Russia, and the United States are the three space superpowers, followed by countries with a substantial space industry, but without their own capacity to launch people into orbit (Europe, Japan,
and India), and then smaller countries able to launch satellites to orbit (Israel, the two parts of partitioned Korea, and Iran). France was
the third country to launch its own satellite, using the desert launch
site of Hammaguir, Algeria. When Europe began to develop its large
rocket through the European Launcher Development Organisation
(ELDO), it turned to another desert site. Because the rocket's first
stage, Blue Streak, was British, it was natural that a British-connected
location should be used – indeed, it was the world's largest launch
range by size. Woomera, in the south Australian desert, was a military site used for nuclear weapons testing, missiles, and sounding
rockets. The European launcher required the digging out of a substantial flame trench and pier, not unlike the Soviet Soyuz, so this
became a big project. It was not rewarded with success, for no ELDO
rocket to orbit was successful. Woomera, though, did witness two
successful satellite launches: an Australian satellite on an American
rocket (WRESAT, 1967) and a British Black Arrow (1971), which put
the scientific satellite Prospero into orbit, but there were no more,
for Britain cancelled both Blue Streak and Black Arrow. Woomera
has now been reclaimed by nature – the fate of several of the early
launch sites.

As France took over leadership of the European space programme,
a new location was sought, the ideal one being French and
European Union territory in French Guiana, close to the equator
in South America and well-suited for launches to 24-hour equatorial orbit. This was a success story. Oceanside Kourou started with

1933

Members of the Group for the Study of Reactive Motion (GIRD) pose with their first liquid-fuel rocket on the outskirts of Moscow, among them the future leader of the Soviet space programme, Sergei Korolev.
Source: National Air and Space Museum, Smithsonian Institution

1936

In the valleys of the San Gabriel Mountains of California, a ragtag group of Caltech researchers test their homemade rockets. The 'Suicide Squad', as they were known on the Caltech campus, lay the foundations of the US' Jet Propulsion Laboratory.
Source: JPL, NASA

1955

Hideo Itokawa, the pioneer of Japanese rocketry, sits at a pencil rocket test stand in Chiba, a western district of Tokyo. Despite its miniature appearance, these small rockets lay the foundations for Japan's space programme.
Source: JAXA

1963

Today, rockets measuring more than 40 m in height overshadow the Satish Dhawan Space Centre, however, it was around 60 years ago when two men pushed a bicycle carrying India's interstellar ambitions to the small Thumba Equatorial Rocket Launching Station.
Source: ISRO

small French rockets and from 1979, became the home of Europe's Ariane rocket, now in its sixth iteration. Although a somewhat remote location to visit, Kourou became known to a worldwide audience because of its steamy jungle, palm trees, Devil's Island off the coast, and historically its penal colony, immortalised on screen by the film *Papillon*. When Europe adopted the Russian Soyuz as its medium-lift launcher, Kourou witnessed the improbable sight of a Russian launch pad being constructed there, with the Soyuz first launched from there in 2011.

The next spacefaring nation was Japan (1970). Beachside sites proving too small and hazardous to onlookers, Japan began construction not of one but two oceanside sites, among the most scenic in the world. The reason for two sites lies in the institutional structure of the Japanese programme. Its early part was developed by chief designer Hideo Itokawa of the Institute for Space and Aeronautical Science (ISAS), who built small, solid fuel sounding rockets and then orbital launchers to put quite small, but nonetheless successful payloads into orbit. He chose Uchinoura, Kagoshima, at the very southern end of the Japanese archipelago. The government programme, developed by the National Space Development Agency (NASDA), developed much larger rockets for satellites for technology, communications, applications, military observations, and to support the International Space Station and for this purpose built its own site not far away on an island off the south coast in Tanegashima. Although NASDA and ISAS subsequently united as the Japan Aerospace Exploration Agency (JAXA), they retain two distinct launch bases, with Tanegashima being the home of the H rocket and Uchinoura that of the Epsilon

small solid-fuel launcher. These sites had the great advantage of bringing the rocket over the sea straight away to head out over the Pacific. The land, though, was quite rugged, requiring substantial carving out of the hills. Perhaps the biggest initial challenge was presented by local fisherman, who feared that the noisy rockets would disturb the fish – or require the disruptive evacuation of their ships downrange of launches. In the end, quite limited launching seasons were negotiated.

Rather like Japan, India's first launch site was a beachside location: Thumba, near the town of Thiruvananthapuram on the south-west coast, was used for sounding rockets. The key consideration was actually a building where small rockets could be assembled. The local bishop and its congregation gave permission for the use of the church for preparing the rockets, which were then transported to the beach by bicycle. Soon, larger sounding rockets supplied by France, the United States, Britain, and the Soviet Union were fired from there and it was formally designated by the United Nations an international equatorial sounding rocket facility. To reach orbit, much larger rockets and a full-scale launch facility were required, leading to the selection of a sand ridge on the east coast, Sriharikota on the Bay of Bengal, north of Thumba's latitude but on the east coast close to the equator, with the nearest city being Chennai. With its combination of lakes, flatness, wildlife, streams, and sand spits, it most closely resembled Cape Canaveral in appearance. Sriharikota saw the launch of India's first satellite on an Indian launcher (1980) and the development of a space programme highly focussed on economic, social, and technological development. New pads were

1956

Qian Xuesen spent five years under restrictions before being deported by the US in 1955 as part of a domestic crackdown against communism. Only one year later he can be seen at a banquet next to Mao Zedong. Xuesen would move on to become the founding father of China's missile and space programme.
Source: China Academy of Launch Vehicle Technology

built for India's ever more powerful launchers: the Polar Satellite Launch Vehicle (PSLV) and the Geosynchronous Launch Vehicle (GSLV). Sriharikota attracted ever more international attention when India sent its first spacecraft, Chandrayaan, to the Moon in 2008 and then its Mars Orbiter to Mars in 2014.

Now come a number of smaller space powers, smaller in the sense that their launch capacity was limited, generally to quite small satellites. Israel was the first, launching its own first satellite in 1988 using a coastal launch site called Palmachim. Israel's satellites have all been for military observation purposes, their most striking feature being that they launch westward towards the Mediterranean. All other countries launch to the east to gain the benefit of the spin of the Earth on its axis, but this is not possible for Israel, for an eastward launch would fly over hostile neighbours who might mistake it for a missile attack. Next came the two parts of Korea. The Democratic People's Republic of Korea (DPRK) announced its first satellite launch in August 1998, a second in February 2009, and a third in April 2009, all unsuccessful and from a basic, gravel-road site in isolated hills to the north-east of the country at Musudan-Ri. Launches were then moved to a more permanent facility, Sohae, in the north-west of the country near the border with China, from where a successful launch finally took place on 12 December 2012. Following the long tradition of overflights of secret cosmodromes, Musudan-Ri and Sohae became the most surveilled, monitored, and photographed launch sites in the world.

The DPRK narrowly beat the Republic of Korea into orbit. The republic sought the help of experienced, veteran cosmodrome builders and rocket and rocket engine makers in Russia to build a coastal site, Naro, on Oenaro Island; provide a rocket, the Korean Satellite Launch Vehicle (KSLV) 1; and engines, the RD-151, which on the third attempt put a science and technology satellite into orbit on 30 January 2013, which was succeeded by a domestic launcher, the KSLV II, in the 2020s.

The most recent member of the space club is Iran, which sent its small Omid satellite into orbit on 4 February 2009 using the small Safir rocket, just able to reach orbit. Iran used one of its military bases in the Semnan area, hence the name applied to the site, and it was, as one might imagine, a basic facility of concrete platforms, gantries, and transporters. Like the DPRK, the site was modernised for a larger rocket, in this case the Simorgh, with permanent facilities and a new name: the Imam Khomeini Launch Centre, with an image of the Imam imprinted on the new launch tower.

The candidate for next country to reach orbit has long been Brazil, which built the coastal Alcântara Space Centre and developed its own launcher, the VLS. The first two launch attempts were unsuccessful, but worse followed on 22 August 2003 when the rocket exploded on the pad during pre-launch preparations, killing 21 people. Alcântara was rebuilt and re-opened as a fully-equipped launch centre. Grounded by the United States, Alcântara has yet to experience the joy of a successful satellite launch.

Although these are the best-known and 'first satellite' launch sites of the large, medium, and small space powers, there are more. In 1961, the United States opened an oceanside launch site up the coast from Cape Canaveral at Wallops Island, Virginia, used initially for the Scout rocket for small scientific satellites,

2009: A model of Iran's Safir rocket, which carried Iran's first satellite Omid ('Hope') into orbit. It accompanies the anniversary celebrations of Iran's 1979 Islamic Revolution on Tehran's Azadi Square.
Source: Picture Alliance / REUTERS / CAREN FIROUZ

Left:
On 15 November 1988, the Energia Buran lifted off on its only flight. Intended for human flight once tested, the shuttle performed its entire mission, including the landing, uncrewed. Afterwards, it was left to deteriorate in the facilities that housed it.
Source: Picture Alliance / dpa / Alexander Mokletsov

but re-opened in recent years for Antares rockets to fly to the International Space Station. On the west coast, a largely military launch site was opened in 1959 at Vandenberg Air Force Base along the hill shoreline of the Pacific coast, once intended to be a second home to the space shuttle. It is to Cape Canaveral what Plesetsk is to Russia's Baikonur: a busy, largely military base. The United States also developed two small military defence sites with infrequent orbital launches: Kodiak, Alaska, and Omelek, Kwajalein Atoll in the Marshall Islands, which have launched 11 and two satellites, respectively. Russia used two military bases for satellite launches using decommissioned Cold War launchers: Yasny, also called Dombravoska, in southern Russia (92 satellites) and Svobodny in the far east (five satellites), now disused. Russia even fired three small satellites into orbit from the submarines Ekaterinberg and Novomoskovsky.

Finally, there have been unusual launch sites. Two were offshore: San Marco and Sea Launch. In 1961, Italy responded to the invitation by the United States to assist European countries in launching satellites. Italy chose to do so by using the small American Scout rocket to launch Italian satellites. Rather than do so from Italy, which would have presented downrange problems, Professor Luigi Broglio of the Italian Space Commission chose a launch site in international waters off the coast of Kenya, ideal for reaching equatorial orbit and with the Indian Ocean as the downrange area. This used a towed oil rig as a launch platform for the Scout and an oil platform as mission control. An Italian satellite, San Marco, was fired from there in 1967, with the site being used for Italian, American, and British launches, nine in all, until 1988.

The second such site was Sea Launch, developed during the heady days of the commercialisation of the Russian space programme with Russian, US, and Ukrainian cooperation. With a similar approach to reaching equatorial orbit, a large oil rig, Odyssey, was towed to the mid-Pacific, with Zenit rockets from Ukraine being ferried there by a large ship, Sea Commander, with a home port at Long Beach, overseen by the Boeing company (US) and the Energiya design bureau (Russia). Its principal function was to send commercially-lucrative communications satellites to 24-hour orbit, with 36 launches from 1999–2014. Launches were suspended when the supply of Zenits ceased. In 2020, Sea Launch was towed across the Pacific for a new life in Slavyanka in the Russian far east. China followed this launch method on 5 June 2019 when the small CZ-11 solid-fuel launcher on its seventh flight was fired from a converted drilling rig, the Tairu, towed by two tugs into the Yellow Sea to launch seven small payloads. This was a successful proof-of-concept test witnessed by paying guests who chartered a ship for a two-day cruise to witness the spectacle.

Some satellites reach orbit without any launch site at all. The idea of launching satellites from aircraft goes back to the early space age, the idea being that an aircraft can do some of the heavy lifting and bring a rocket with a satellite up through the heavy layers of the atmosphere, where a smaller rocket can take over and bring it into orbit. Air launch is not as easy as it looks, and early attempts had their share of difficulties. Air launch was first achieved on 5 April 1990 when a Boeing B-52 brought a Pegasus rocket up to 12,000 m, released it, and the Pegasus fired its engine to bring a small military satellite into orbit, the first of six such launches. Later, the Lockheed L-1011 Tristar, with greater lifting capacity (350 kg), was used. Aircraft had the advantage that they could launch satellites from any location with an ordinary aircraft runway, so typical locations were Edwards Air Force Base, California; Cape Canaveral; Kwajalein Atoll; Wallops Island, Virginia; and the Canary Islands (Spain). By spring 2022, 108 satellites had been air-launched. A new company joined in 2020, Virgin, with its Boeing 747 Cosmic Girl successfully lifting a satellite to orbit the following year.

In a variation of this theme, the International Space Station (ISS) has been used as a launch pad for satellites. The practice actually dates back to the Soviet space programme, but it became a feature of the world space industry with the ISS when it became economical to use cargo craft to bring very small satellites up to its altitude (400 km) and then launch them. Such satellites could be hand-launched (Russian practice) or use a specially-designed dispenser on the Japanese module (Kibo). By autumn 2021, 533 satellites, mainly micro-satellites, had been launched from the ISS and earlier space stations. Small satellites have even been launched from lunar orbit, starting with the ejection of a small scientific satellite from Apollo 15 on 4 August 1971.

Now comes a new word to add to the generic term 'launch site' (and in Russian, 'cosmodrome'): 'spaceport'. With the advent of commercialisation and space tourism, a host of new locations are planned and for these, the term 'spaceport' has become popular. Some spaceports do not intend to reach orbit, like the Virgin Galactic spaceport that will make sub-orbit flights, but others do, like the planned Scottish spaceport in Sutherland and its companion site at Malbusca, Santa Maria in the Azores. So far, only

Airport or James Bond villain's den? Neither. Contrary to the vertical launch platforms of other spaceports, Virgin Galactic's Spaceport America stands out of the Sierra County desert in New Mexico through its deliberately futuristic architecture.
Source: Virgin Galactic Press

one spaceport has been successful: Mahia in New Zealand, run by an American-New Zealander team. This is a story with many future chapters.

The launch sites of the world are a microcosm of space history and the efforts of different countries, separately and together, to reach space. They tell stories of the Cold War, cooperation, achievement, tragedy, and disaster. The different locations – desert, sandy seaside, hilly coast, tundra, mountains, valleys, and jungle – are stories of geography, climate, and politics. Some are well known to the world, others obscure. Some are well-trampled by tourists, while others remain remote, some still secret. Some of the busiest (Plesetsk, Vandenberg) are little known, while others like Cape Canaveral have been an indelible part of popular culture and the vernacular since the late 1960s. At one extreme, some are the epitome of modernity – Vostochny, Wenchang, Kourou – while wildlife scampers over the rusting remains of historic sites, such as Woomera and Kapustin Yar.

Their architectures began as a cement stand set in a forest (Peenemünde) and evolved to ever-more complex gantries, service towers, crawlers, mobile towers, assembly and integration buildings, and industrial and processing facilities. They were served by combinations of roads, railways, and aircraft runways. The trajectory from simplicity to sophistication is evident – although the early rocketeers would have said that there was nothing simple about them. Launch sites are intensely utilitarian: they are designed as efficient, connected structures that enable rockets to safely reach space. Wernher von Braun was once confronted by a critic who told him that his creations were 'not very beautiful'.

'What works is beautiful,' he retorted, and the architecture of launch sites should be judged on that basis. Nevertheless, it has produced some unforgettable and iconic architectural structures, of which the Vehicle Assembly Building (VAB) at Cape Canaveral must top the list on account of its size, function, domination of the landscape, and statement of the ambition of conquering the Moon. It is little surprise that it has been flattered by its subsequent imitations in Vandenberg, Jiuquan, and Wenchang. 'What works is beautiful' is true of the first launch pad of the modern space age, the subsequently named Gagarinsky Start, pad 1 at Baikonur, with its rugged excavated flame trench, launch site on pillars, complex installation cabin underneath the rocket tail, and elegant system of weights and tulip-shaped arms that fall back as thrust exceeds gravity. It was a system so effective that it, too, was imitated at Plesetsk at the time and at Kourou and Vostochny 50 years later.

The purpose of this book is to tell the architecture of launch sites in an accessible way. Although technical details are rightly given – to convey size, shape, idea, purpose, and function – it is not intended as a purely technical description. Rather, it situates launch sites in their context, history, evolution, development, narrative, changing use, and future prospects, how they work, their different elements, and how they make up a part of their country's space programme and the world space endeavour as a whole. It is not a directory of every launch site ever built and should not be seen as such. It does include every significant launch site of the space age, past and present, not least those expected to make an ever-greater impact in the future (e.g., Wenchang) or those that

With a different kind of showmanship, SpaceX shows off its immense technical accomplishment of a reusable Falcon 9 first stage booster landing at Vandenberg Air Force Base in California.
Source: Picture Alliance / ZUMAPRESS.com / Gene Blevins

have the prospect of doing so (Alcântara). It does include some sites that are no longer in use but played an important part of the history and architecture (Peenemünde, Hammaguir, Woomera, Kapustin Yar). It includes at least one from every currently spacefaring nation, including newcomers (DPRK, Rep. Korea, Iran), even if they have witnessed few launches and about which little may be known. Some minor, decommissioned, or military sites are not included because of their limited significance; nor are there lists of air-launched satellites nor those ejected from space stations. Sites such as Spadeadam, Westcott, and the Isle of Wight in Britain were used only for static testing of rocket engines: launches were not conducted from there and they are thus also omitted. The two maritime sites, San Marco and Sea Launch, are distinctive, lateral-thinking, proven ways of launching satellites with a distinctive architecture, so they are included. In the text on launch sites, those of India were written by Gurbir Singh and the others by Brian Harvey.

Finally, a new, hitherto uncovered, unacknowledged aspect of launch site research has begun to open up in recent years, namely their social histories and consequences. Several launch sites – and other rocket facilities – were built at the expense and pain of indigenous peoples who were displaced. This is now under investigation by the scholar Asif Siddiqi and the reader is recommended *Departure Gates: Postcolonial Histories of Space on Earth* (MIT Press, due 2022 – 2023).

Acknowledgements
We wish to thank and acknowledge the following for their assistance: Anatoly Zak; Andy Thomas for Woomera photographs and Morgan Bailey, Rocket Lab, for Mahia photographs. Figures of satellite launches are taken from the United Nations Office for Outer Space Activities (UNOOSA) register.

Further reading
Readers are recommended this excellent directory (2005) by Stephen Strom: Strom, Stephen, *International Launch Site Guide*. 2nd edition. (El Segundo, California, and Reston, Virginia: Aerospace Press & American Institute of Aeronautics and Astronautics, 2005).
A current directory of launch sites may be found at <https://www.daviddarling.info/encyclopedia/L/launch_sites.html>
Launch sites are one of the topics reported by Mark Wade's Encyclopedia Astronautica: <www.astronautix.com>
Launches to space and interceptions have taken place from aircraft, ships, and submarines. <http://www.spacetoday.org/Rockets/Plowshares/Submarine.html>, <https://en.wikipedia.org/wiki/Operation_Burnt_Frost>
Around the world in seven spaceports: <https://storymaps.esri.com/stories/2019/spaceports/>
Gunter Krebs maintains an interactive map of launch sites that includes orbital, suborbital, and military launch sites. <https://space.skyrocket.de/directories/launchsites.htm>
The following Wikipedia entry lists a sortable table of launch sites with characteristics including location, coordinates, operational date, number of launches, etc. <https://en.wikipedia.org/wiki/List_of_rocket_launch_sites>
NASA launch sites: <https://www.nasa.gov/centers/kennedy/launchingrockets/sites.html>
Centre for Strategic and International Studies maintains a map of spaceports of the world. <https://aerospace.csis.org/data/spaceports-of-the-world/>
Spaceflightnow maintains a calendar of all upcoming launches and frequently broadcasts them live online. <https://spaceflightnow.com/launch-schedule/>

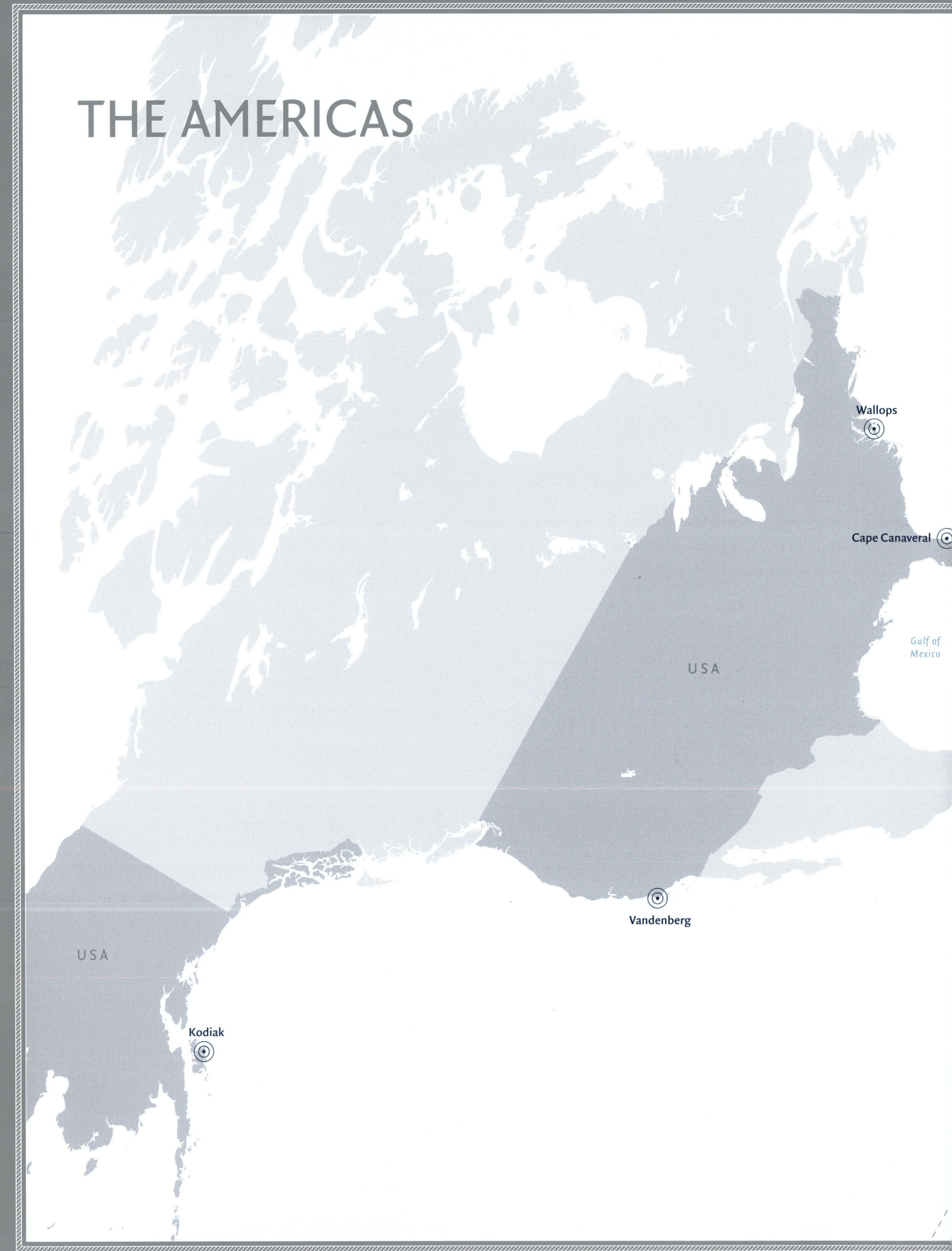

THE AMERICAS

Wallops

Cape Canaveral

Gulf of
Mexico

USA

Vandenberg

USA

Kodiak

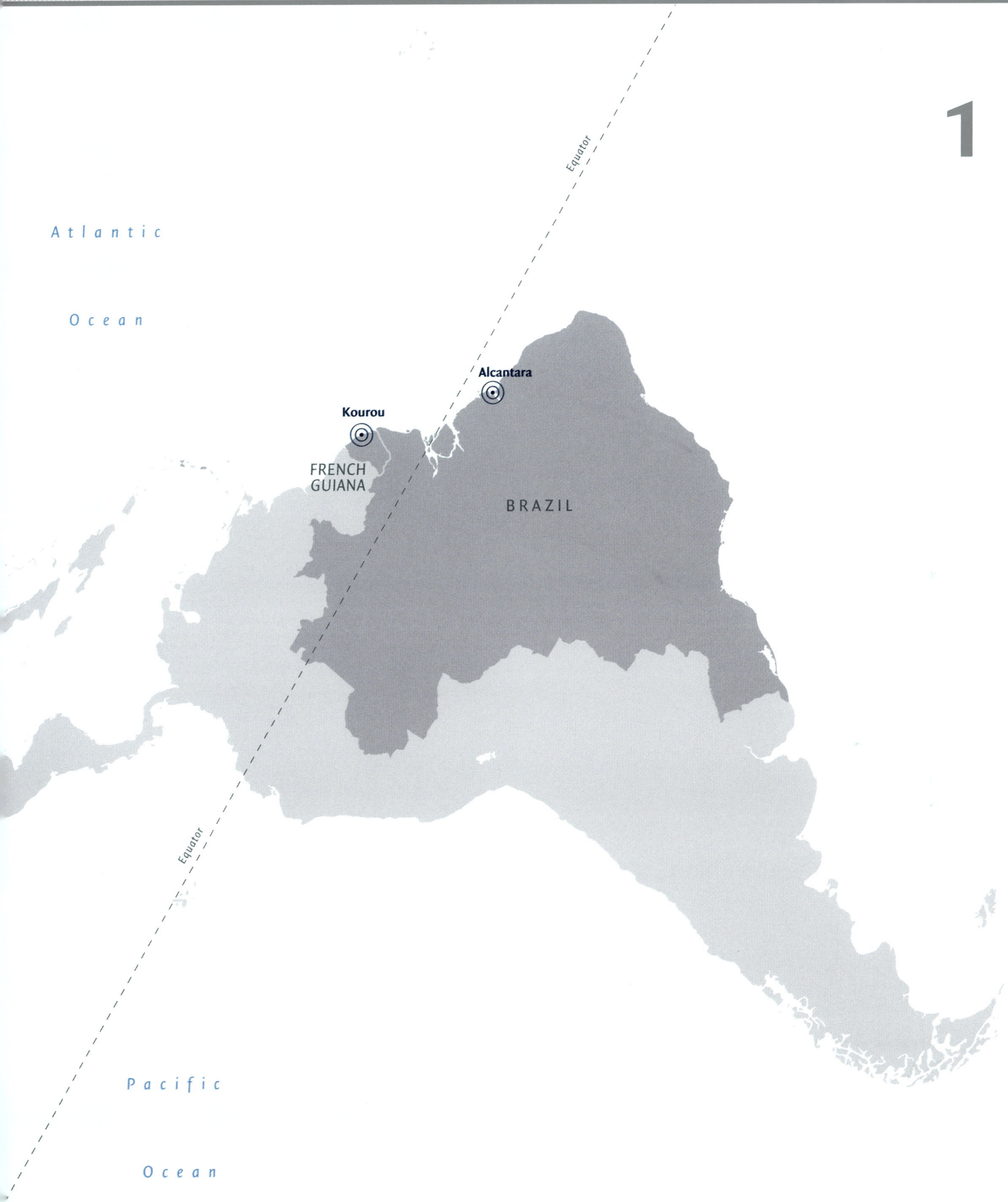

Atlantic

Ocean

Equator

Alcantara

Kourou

FRENCH
GUIANA

BRAZIL

Equator

Pacific

Ocean

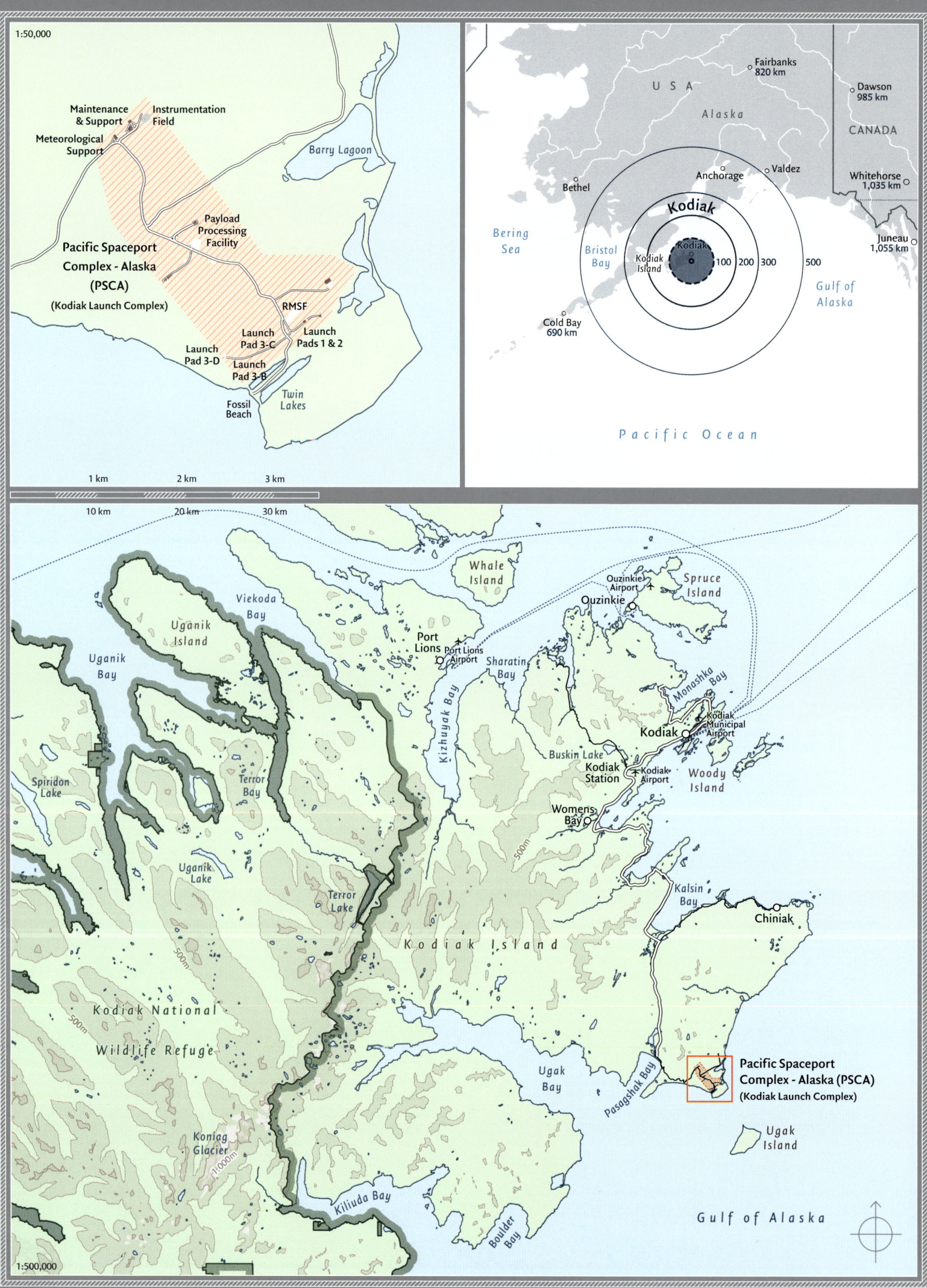

1:50,000

Maintenance & Support

Instrumentation Field

Meteorological Support

Barry Lagoon

Pacific Spaceport Complex - Alaska (PSCA)

(Kodiak Launch Complex)

Payload Processing Facility

RMSF

Launch Pad 3-C

Launch Pads 1 & 2

Launch Pad 3-D

Launch Pad 3-B

Fossil Beach

Twin Lakes

1 km   2 km   3 km

Fairbanks 820 km

Dawson 985 km

USA

Alaska

CANADA

Bethel

Anchorage

Valdez

Whitehorse 1,035 km

Bering Sea

Kodiak

Bristol Bay

Kodiak Island

Kodiak

100 200 300   500

Juneau 1,055 km

Gulf of Alaska

Cold Bay 690 km

Pacific Ocean

10 km   20 km   30 km

Whale Island

Ouzinkie Airport

Spruce Island

Ouzinkie

Viekoda Bay

Uganik Island

Port Lions

Port Lions Airport

Sharatin Bay

Monashka Bay

Uganik Bay

Kizhuyak Bay

Kodiak

Kodiak Municipal Airport

Spiridon Lake

Terror Bay

Buskin Lake

Kodiak Station

Kodiak Airport

Woody Island

Uganik Lake

Terror Lake

Womens Bay

Kodiak Island

Kalsin Bay

Chiniak

Kodiak National

500m

Wildlife Refuge

500m

Ugak Bay

Pasagshak Bay

Pacific Spaceport Complex - Alaska (PSCA)

(Kodiak Launch Complex)

Koniag Glacier

1,000m

Ugak Island

Kiliuda Bay

Boulder Bay

Gulf of Alaska

1:500,000

ALASKA'S ISLAND SPACEPORT

Name	Pacific Spaceport Complex	**Coordinates**	57° N, 152° W

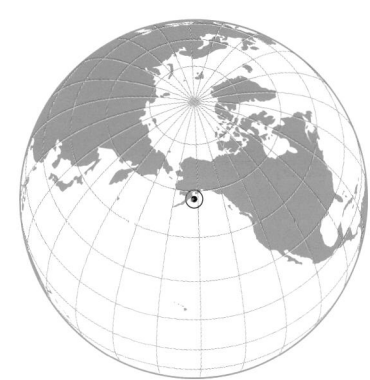

Name	**Coordinates**
Pacific Spaceport Complex	57° N, 152° W
Location	**Time zone**
Alaska, United States	GMT-9
Owner / Operator	**Launches**
Alaska Aerospace Corp	12 satellites
Elevation	**Completion**
1 m	2001

Kodiak opened in 1998 as a public-private government-commercial initiative of the state of Alaska. Its formal title is now the Pacific Spaceport Complex Alaska. Its debut launch was the Kodiak Star using the small solid-fuel Athena 1 rocket on 30 September 2001 to orbit four payloads. This was followed by eight satellites for the US Air Force on the solid-fuel Minotaur IV (2010) and another Air Force mission with a single satellite the following year. Orbital launch attempts resumed with the 12-metre-tall liquid-fuel Astra rocket. Astra is a startup based in Alameda, California, and Kodiak's anchor commercial operator, competing for military and commercial orbital payloads up to 150 kg in the small launcher market. Its first two satellites were put into orbit over 2021 – 2022. The launch centre is located on 15 km² of state land. The site is on a promontory in Pasagshak on the east side of the island, the best reference point being offshore Ugak Island. It has two main launch pads – one for orbital and the other for sub-orbital missions – and several flat concrete pads. It has facilities for payload and integration processing, storage, and solid and liquid fuelling. There is a 17-floor enclosed movable gantry, 120-place launch control centre, and tracking services. Southward launches have a wide range of azimuths available to them over the Pacific – Kodiak's biggest selling point. There is a permanent staff of 30 on site. There are ambitious plans to expand the launch rate, develop visitor facilities, and host stratospheric balloon launches. Kodiak Island (capital: Kodiak), home of the Alutiiq people and part of Russia until 1867, is a large island, bigger than Cyprus, off southern Alaska at the beginning of the Aleutian chain of islands that stretch south-westward into the Pacific.

The launch tower and launch vehicle integration building housed the first government launch in 1998 and first commercial launch in 2018 on the green coastline of Kodiak Island.
Source: NASA

August 2014: The launch pad and neighbouring facilities are left destroyed due to the launch termination of a rocket during the beginning of its flight.
Source: Alaska Aerospace Corporation

Astra, a relatively small satellite launch
company, has been launching from the
spaceport in Kodiak since late 2020.
Source: Alaska Aerospace Corporation

Next double page:
Kodiak launch centre in nature.
Source: Shutterstock / Shchekoldin Mikhai

Half the island is a wildlife reserve, with its own type of bear (the
Kodiak bear), and it is famous for its salmon. Kodiak's out-of-the-
way location means that it is not well-known, but visitors may
explore the site's trails to spot whales, bison, and eagles. When
launches take place, hundreds of visitors and residents gather to
watch from the surrounding hills.

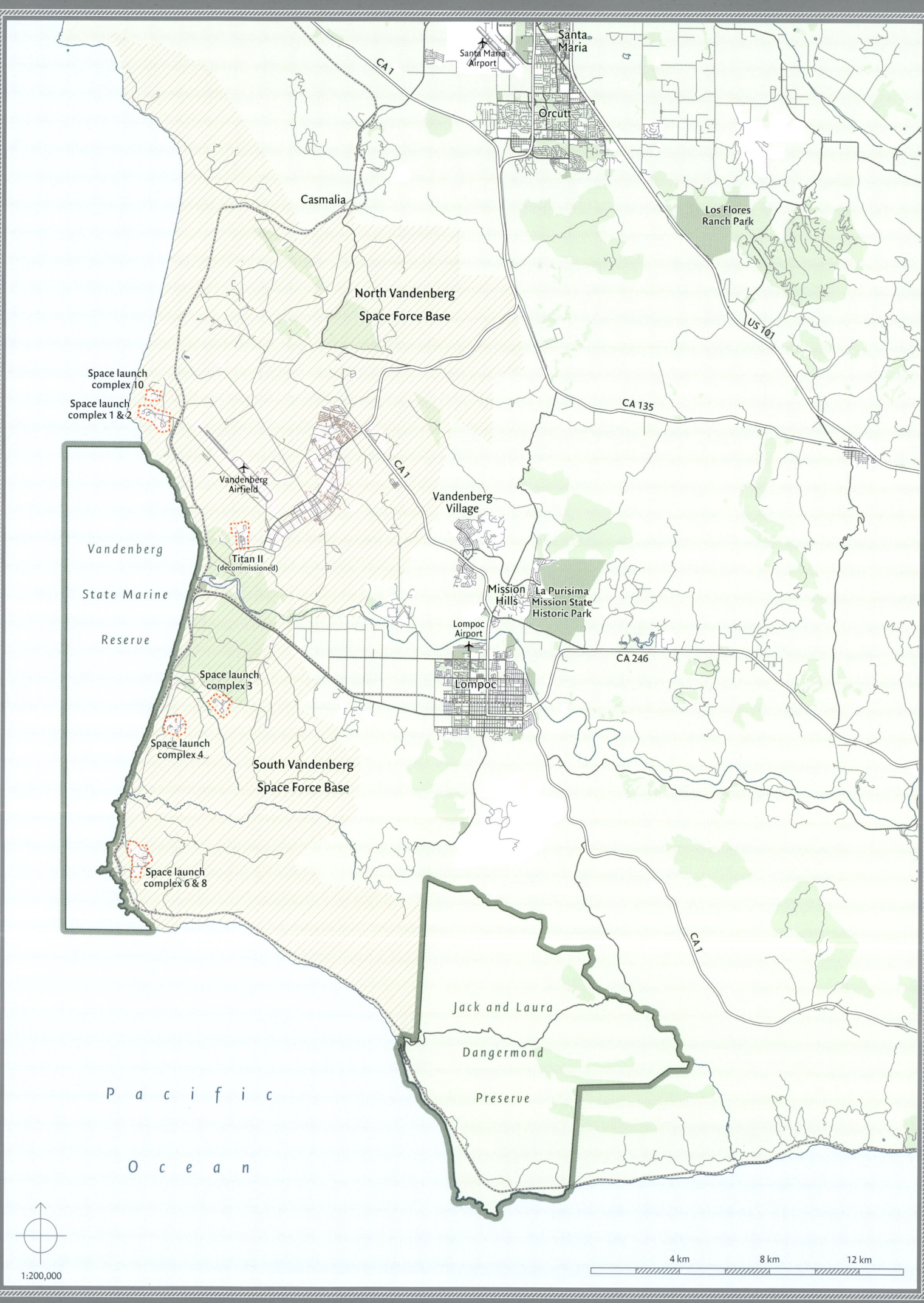

Santa Maria
Santa Maria Airport
Orcutt
CA 1
US 101
Los Flores Ranch Park
Casmalia
North Vandenberg Space Force Base
CA 135
Space launch complex 10
Space launch complex 1 & 2
CA 1
Vandenberg Airfield
Vandenberg Village
Titan II (decommissioned)
Vandenberg State Marine Reserve
Mission Hills
La Purisima Mission State Historic Park
Lompoc Airport
CA 246
Space launch complex 3
Lompoc
Space launch complex 4
South Vandenberg Space Force Base
Space launch complex 6 & 8
CA 1
Jack and Laura Dangermond Preserve
Pacific Ocean

1:200,000

4 km   8 km   12 km

BUSY WEST COAST MILITARY BASE

Name		Coordinates
Vandenberg Space Force Base		34.4°N, 120.4°W
Location		Time zone
California, United States		GMT-7
Owner / Operator		Launches
United States Space Force		1,280 satellites
Elevation		Completion
45 m		1959

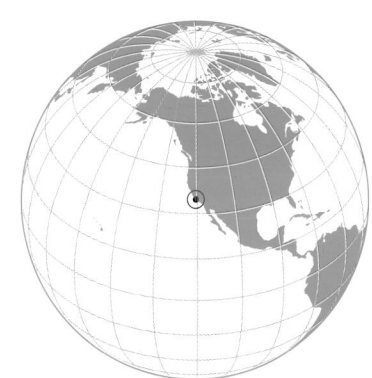

Vandenberg on the west coast is hardly known compared to Cape Canaveral (qv) on the east coast. It is little realised now that a space shuttle base was once built there. It is the principal military base for the United States, mainly but not entirely used for military satellite launches. By contrast, Cape Canaveral is principally used for civilian and scientific launches, although military rockets have also flown from there. Vandenberg's location enables polar launches over the sea. The first polar launch from Cape Canaveral at the beginning of the space age dropped debris on Cuba, fatally injuring a cow, so it was important to find a location without undiplomatic consequences. Vandenberg had the added attraction of remoteness, but its downside was that moist air from the Pacific meeting its rolling hills and canyons was inclined to turn into fog.

The site's origins go back further. In the 1950s, the United States Air Force (USAF) sought a training base for crews launching its missiles from Cape Canaveral, settling on Camp Cooke, an army camp on a hilly location on the Pacific coast 251 km from Los Angeles and 450 km from San Francisco. It was just north of Point Arguello along the Santa Ynez River. The army bought 35,000 ha of land around the town of Lompoc in 1941, its varied terrain making it an ideal training ground for infantry and tanks. During the Second World War, it also became a camp for German and Italian prisoners. It was deactivated after the Korean War, with some barracks being turned over to the prison service in Lompoc. In 1958, this training base was named Vandenberg Air Force Base (AFB) after the former Air Force Chief of Staff General Hoyt Vandenberg, but it is also called the Western Test Range (WTR).

Left:
The Mate-Demate Device designed and used to lift a space shuttle orbiter onto and off the back of a Shuttle Carrier Aircraft. In total four Mate-Demate Devices were built across the US, including at Kennedy Space Centre, to make possible the transport of the reusable shuttle across the United States.
Source: NASA

Right:
2018: A SpaceX Falcon 9 rocket carrying the Iridium-5 satellite blasts off its launch pad at Vandenberg Air Force Base.
Source: Picture Alliance / ZUMAPRESS.com / SpaceX

Although an extensive military facility, the coastal Pacific railway runs north-south through the launch site and in the middle, there is an east-west corridor for civilians to reach the beaches at the village of Surf. This divides the base in two: North and South Vandenberg.

Vandenberg quickly became a launch base in its own right. The first missile was fired from Vandenberg on 16 December 1958 and the first orbital launch, Discoverer 1, launched into polar orbit on 28 February 1959. Disguised as a civilian mission, it was for military photoreconnaissance and the majority of American military satellites have subsequently been launched from Vandenberg. At one stage, Vandenberg had 53 launch pads, mainly military, including Minuteman silos as part of the live nuclear deterrent. These pads were used by a wide range of military missiles, such as Titan, Atlas, Thor, and Scout. Royal Air Force crews for Britain's Thor rocket nuclear deterrent trained there.

Space Launch Complex 5 (SLC-5 or 'Slick 5' in the argot) was famously used for the main American scientific programme, Explorer, starting with Explorer 19 launched from there in 1963 and was ideal for its polar orbiting missions. SLC-5 was the base of its main early launcher – the small, solid-fuel, four stage Scout rocket – launched from there 69 times until 1994. The Air Force attempted to develop a military version, the Blue Scout, for small military payloads. There were numerous failures, causing the Air Force to rush to retrieve its payloads from the ocean off the coast before anyone else did and the project was abandoned. SLC-5 was then decommissioned and stripped. Later Explorers also used Vandenberg on Delta rockets.

Vandenberg was intended for crewed flight twice. In the 1960s, Space Launch Complex 6 (SLC-6 or 'Slick Six') was built with two pads for the Titan III-based Manned Orbiting Laboratory (MOL). Vandenberg was chosen because it required a polar orbit for its purpose of military reconnaissance; conversely there was the danger that launching out of Cape Canaveral would risk descending on top of Cuba and falling into enemy hands. SLC-6 was a huge construction project, with the US Army Corps of Engineers shifting 1.1m m³ of earthworks to build a launch pad, mobile service tower, umbilical tower, and launch control and service building. It was completed in 1968. MOL required a further extension of the base to 41,000 ha, making it the third-largest Air Force Base anywhere, after Eglin and Edwards. The following year, the project was cancelled and SLC-6 was mothballed.

Several years later, the Air Force joined the shuttle programme and SLC-6 was re-built as a shuttle base, enabling the shuttle to fly into polar orbit for reconnaissance missions, with at least one shuttle to be stationed there. This was a €4bn project that involved construction of a Shuttle Assembly Building (SAB), access tower, launch mount, mobile service tower, and a pad with three exhaust ducts, all in a much more compact space than the facilities at Cape Canaveral. The shuttle was stacked directly on the pad, protected by the launch tower, mobile service structure, and payload checkout facility, rather than in the assembly building and then brought to the pad, which was the procedure at Cape Canaveral. An escape system was installed to enable the quick evacuation of the shuttle cabin before launch. The runway was extended to 4.5 km with a Precision Approach Path Indicator lighting system to guide

The US Air Force pleaded to the Senate
for funding to construct site 6, justifying
a need for placing high-priority military
payloads into polar orbit. Earlier, it
intended to launch the Manned Orbiting
Laboratory reconnaissance space station
until the programme was cancelled.
Source: NASA

Top:
1985: The Enterprise shuttle sits atop
SLC-6 for fitting checks. SLC-6 was
intended for military launches to polar
orbit from the west coast for secret
surveillance satellites.
Source: US Air Force Archive

Right:
A satellite is transported across
Vandenberg Air Force Base to Space
Launch Complex 2 to be fitted to a
Delta II rocket for launch.
Source: NASA / Randy Beaudoin

Wildfires burn in the background of the Atlas 5
rocket pad in September 2016.
Source: Santa Barbara County Fire Department

Right:
United Launch Alliance's Delta IV-Heavy rocket
launches from Space Launch Complex-6. This
was the second Delta IV-Heavy launch for
Vandenberg and the largest to ever launch
from the west coast of the United States.
Source: U.S. Air Force photo / Joe Davila

2019: A 69-metre-tall Falcon 9 stands ready for
lift-off on Space Launch Complex 4-East.
Source: SpaceX

in returning shuttles. An Orbiter Maintenance and Checkout
Facility (OMCF) was built. Tanks were installed for liquid oxygen
and hydrogen fuels, as were storage facilities for the Solid Rocket
Boosters. To deliver the shuttle from its Boeing 747 carrier, a
35 km route was cut between the airport and the OMCF. The facilities were almost completed in 1986, with the first shuttle mission from Vandenberg scheduled for 15 October that year, when
Challenger exploded at Cape Canaveral that January. The idea of
basing the shuttle at Vandenberg was abandoned and the expensive SLC-6 facilities were again mothballed. In the new century,
SLC-6 was re-opened for the Delta IV rocket. The runway came
into use too, for in 2010 it began to be used for the autonomous
landings of the unpiloted Air Force X-37B orbital space plane.
Several prominent civilian space missions have been launched
from Vandenberg, notably weather satellites into polar orbit. The
first interplanetary launch from Vandenberg – due to congestion
at Cape Canaveral – was InSight on an Atlas V on 5 May 2018. It
was sent to Mars, where it landed that November. Because there
is no public viewing in the base itself, NASA made arrangements
for enthusiasts to view the launch 15 km from the launch site at
Lompoc City Airport and St Mary's Episcopal Church.
Vandenberg has had no fatal accidents, the worst event being the
explosion of a Titan IV some 16 seconds after lift-off. The downward blast greatly damaged SLC-4, starting fires and threatening launch controllers. In 2011, SLC-4 was taken over by SpaceX,
which completely rebuilt the area for its Falcon 9 launcher, which

first flew from there in 2013. On its second launch, SpaceX demon-
strated its spectacular ability to recover the first stage of a rocket
when it returned on its tail to land on a cement ground pad at the
base.
Vandenberg is not open to visitors like Cape Canaveral. The only
public views of the area are from the Amtrak train as it passes
56 km through the base and from Surf beach (warning: this beach
is also popular with great white sharks). The coast road turns inland to avoid the base. Most of the land area of the base therefore
serves wildlife and is considered a high-quality natural environment, protecting endangered species from snowy plovers to wa-
tercress. There is provision for visitors through the public affairs
office to the Space and Missile Heritage Centre at SLC-10, which
tells the history of the base, runs exhibitions, and displays historical items. Having said that, night launches can sometimes be seen
from as far away as Los Angeles. About 18,000 people work at the
base, with most commuting from Santa Maria, Santa Barbara, and
San Luis Obispo, and only about 3,000 living in the settlements
close to the base itself. Vandenberg currently has six active launch
sites: 2, 3, 4, 6, 8, and 576. These are now used by the Firefly commercial rocket, Atlas V, SpaceX Falcon, Delta IV, Minotaur, and
Taurus. In practice, the Atlas V and Falcon are the pads likely to
see the most use. The American military space budget has historically been larger than that of NASA and considering the expanding US commitment to the use of space for military purposes, Vandenberg's long-term future is secure.

1965: US Air Force engineers tending to a strategic missile based at the Vandenberg Air Force Base.
Source: USAF Archive

Right: The Space Launch Complex 3 was originally designed for Missile Defence Alarm System launches throughout the early 1960s. Later it would be used for re-entry vehicle tests.
Source: Library of Congress

THE WORLD'S BEST-KNOWN LAUNCH CENTRE

Name
Cape Canaveral Space Force
Station, Kennedy Space Centre

Location
Florida, United States

Owner / Operator
National Aeronautics and
Space Administration

Elevation
7 m

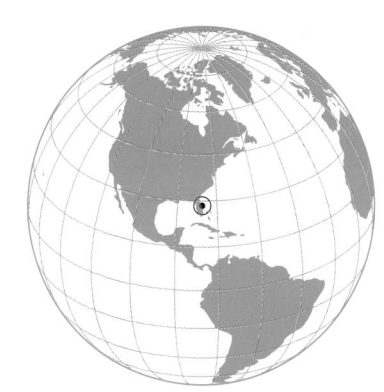

Coordinates
28.5°N, 80.5°W

Time zone
GMT-4

Launches
3655 satellites

Completion
1958

Cape Canaveral has the distinction of being the best-known launch centre in the world. This is because it was the departure point for the Apollo 11 mission to the Moon in 1969. The suitability of the Florida coastline for journeys to the Moon was recognised by science fiction writer Jules Verne, who in 1865 selected it for his travellers' voyage to the Moon. In 1963, it was renamed 'Cape Kennedy' after President John F. Kennedy, who inspired the Moon landing, but this was not popular locally and it reverted to Cape Canaveral 10 years later. To this day, 'Cape Canaveral' comprises the Space Force Station (formerly the Air Force Station; military), and the 570 km² Kennedy Space Centre (civilian). Although best-known for the space programme, Cape Canaveral's history dates to the Spanish conquest (Cabo Cañareal), the lighthouse (1848), the Air Force Station (1949), and the port (1950). It

is a large sand spit on low ground 3 m above sea level, linked by causeways, It has extensive lagoons (Banana River, Indian River) populated by alligators and endangered manatees, beach mice, indigo snakes, sea turtles, and bobcats. From the beginning of the space programme, it entered popular culture due to televised rocket launches and became a destination for holidaymakers and rocket enthusiasts alike, with one million people travelling there to watch the Apollo 11 launch.

Cape Canaveral was originally a missile testing station, with a German A-4 launched from there on 24 July 1950. It had the advantage of ease of access, low population density, launches heading immediately out to sea, and downrange tracking stations and it was also the east coast location closest to the equator. In terms of negatives, launches to polar orbit would bring rockets over inhabited

Left and bottom:
Early A-4/V-2 'Bumper' test
launches conducted on the first
launch pads at Cape Canaveral.
Source: NASA

Throughout the early 1960s, a large 'S' marked the historic hangar that housed astronaut training, crew quarters, and early spacecraft processing in the early stages of America's manned space programme.
Source: NASA

John H. Glenn, Jr. leaves NASA's Hangar S to become the first American to orbit Earth. Today the hangar has fallen prey to the elements and only exists as a physical landmark to America's cradle of space ambition.
Source: NASA

areas (e.g., Cuba, Newfoundland), while the weather was imperfect, being hot, humid and corrosive, changeable, stormy, and even hurricane-prone during the hurricane season. The first United States satellite, Explorer 1, was launched from there in January 1958 and the first robotic Moon probes launched that autumn. Cape Canaveral provided a visual demonstration of the evolution of launch architecture. The first launch sites were little more than concrete pads and, at a safe distance, protective bunkers. As more and more missiles and rockets came to be tested there, new pads were constructed northward up the coastline, each with a numerical designator (pad 1, 2, 3 etc.). These required ever more complex structures, generally called gantries, with so many being built that the landscape along the shoreline was called 'missile row'. Pad 5 was the location of the first American crewed spaceflight: Alan Shepard (1961) launched on an A-4 derived Redstone. This was a cement pad with a cherry-picker to evacuate the cabin in an emergency, with an adjacent preparation area (Hangar S) and mission control centre (Mercury Control). John Glenn (1962) flew on an Atlas rocket from pad 14, which had a large, red A-shaped gantry rolled back before lift-off. Crewed launches moved in 1965 to pad 19 (Gemini, Titan II) and in 1968 to pad 34 (Apollo 7, Saturn IB). From 1965, mission control moved to Houston, Texas. Other famous pads were 17 (Thor, Delta) and 41 (Titan III, Atlas V). The programme to land an astronaut on the Moon involved the biggest single expansion of Cape Canaveral, almost doubling its size. It was designed by famous architect Charles Luckman. A large area to the north of missile row, Merritt Island, was cleared to make way for a double launch pad (39A and B) for the enormous 111-metre-

tall Saturn V rocket, a gravel crawlerway, the Vehicle Assembly Building (VAB), the launch control centre, and a canal to bring in the barges with the huge stages ferried in from Louisiana. At the time, this was the biggest construction project on the planet. The VAB measured 189 m in length by 125 m in width and was 127 m tall. The VAB became one of the iconic architectural structures: simple and huge, the big black 139-metre-tall doors opened into a cavernous interior where four Saturn Vs could be assembled at a time and inspected at multiple levels – so tall that clouds could form within but for air-conditioning. At 3.7m m³ volume, it is still the largest single-floor building in the world, with the world's largest doors and it is the model for launch centres worldwide.

The 5,400-tonne Saturns were brought the 5 km to the pad on one of three 136-metre-tall mobile launch platforms atop a huge crawler moving at 1.4 km an hour, up a ramp to the pad with an umbilical tower, flame deflectors, and water suppression systems. On the pad, the top walkway brought astronauts to their cabin. At lift-off, the arms supplying umbilical power swing back simultaneously. The launch control centre was bright, airy, and often sun-drenched as rockets took off. The moment the Saturn cleared the tower, control passed to mission control in Houston, Texas. A news centre opened in time for the first Saturn V launch in 1967.

After the last Apollo mission (Apollo 18, 1975), Cape Canaveral was suddenly a quiet place, as the space shuttle was being made ready. The VAB was re-purposed for the shuttle. A new 4.7-km runway, one of the world's longest, was built to the north-west of the VAB for returning shuttles, with the first such landing taking place in 1984. Once landed, the shuttle was safed and towed to

Left:
Fourteen years after the first rocket test and five years before the Moon landing, Cape Canaveral had already taken over the entire peninsular. The north-facing aerial view of 'Missile Row' has not changed significantly, besides its launch pad upgrades since the picture was taken in 1964.
Source: NASA

Right and bottom:
The Saturn V rocket and its distinct red mobile launch tower atop a huge crawler-transporter rolling out of Kennedy Space Centre's Vehicle Assembly Building to begin its trek to the launch pad.
Source: NASA

Left:
Construction of the Launch
Control Centre adjacent to the
Vehicle Assembly Building.
Source: NASA

Right:
Enormous forces pull on the steel
ropes as the Saturn V stage floats
through the assembly building.
The red protective covering of
the five rocket engines adding to
the distinct red and white NASA
appearance.
Source: NASA

Orbiter Processing Facilities built beside the VAB to prepare for future missions. Pad 39 was redesigned: the crawler was still used, but the pad now had a shorter fixed and rotating service structure. An emergency evacuation system was installed with a basket, slidewire, bunker, and armoured car. Cape Canaveral saw the first shuttle launch (Columbia, 1981) and last landing (Atlantis, 2011). Sadly, it witnessed the Apollo 1 fire (pad 34, 1967) and the loss of Challenger (1986), and family, friends, and well-wishers waited in vain there at the runway for the return of Columbia (2003). It was intended to base a military shuttle at Vandenberg Air Force Base (qv), a plan abandoned in 1986, but Cape Canaveral was used for classified missions operated under blackout conditions, except for the launch and landing, which were hard to hide in any case.

The retirement of the shuttle in 2011 after 135 missions meant another period of relative quiet while preparations went ahead for the introduction of new spacecraft and rockets. Pad 39A was made available for the commercial SpaceX company to launch the Falcon rocket flying to the International Space Station, while Boeing used the Atlas V from pad 41 for its Starliner. On 30 May 2020, Americans returned to space from Cape Canaveral with the launch from pad 39A of SpaceX's Crew Dragon.

The big new project was the Space Launch System (SLS), a large new Saturn V class rocket to return Americans to the Moon in the 2020s from a refurbished pad 39B. Although more than 50 years old, the Apollo period facilities were as busy as ever and the SLS will be assembled inside the still-operational VAB. It was a different story for some of the older facilities. The old gantries of missile row had rusted and rotted in the humidity, so several were taken down. The tidy-up included the controversial 2016 demolition of the old Mercury mission control, though John Glenn's pad 14 was restored.

Originally, industrial facilities were constructed inland from missile row on the east side of Banana River: solid fuel propellant, liquid fuel storage, engine testing, spin testing, integration building, sterilisation centre, and equipment storage. In the 1960s, a causeway ('NASA Causeway') was built across the river to a new industrial area 8 km south of the VAB. A key element was the Manned Spacecraft Operations Building, now the Neil Armstrong Operations and Checkout Building. This was originally used as astronaut quarters and for altitude testing, then in the 1980s for the Spacelab module on the shuttle, in the 1990s for modules for the International Space Station (Space Station Processing Facility), and in the 2020s for the Orion spacecraft and Artemis programme to return to the Moon. The present industrial area includes buildings for processing payloads, fuels, and hazardous materials.

Although crewed flights have attracted the most popular interest, throughout this period, Cape Canaveral has been and continues to be used for scientific, applications, military, and deep space missions using the full range of the launcher fleet: Atlas, Titan, Scout, Thor, and Delta. Most of the robotic missions that conquered the solar system started in Cape Canaveral.

Cape Canaveral has a well-established visitor centre with static displays of rockets, bus tours of historic spacecraft, visits to pads (launches permitting), and stops at the Cape Canaveral Air Force and Space Museum and the memorial to fallen astronauts. Visitors are welcome to the wildlife refuge. From the time of the first crewed launches, extensive facilities were provided for press and then for the often huge crowds that come to watch the launches.

8 July 2011: Launch control watching the final flight of the space shuttle programme concluding with the lift-off of Atlantis.
Source: NASA

16 July 1969: Some 42 years earlier, another generation of NASA engineers look to Launch Complex 39A to watch Apollo 11 lift-off on its historic lunar mission.
Source: NASA

2017: SpaceX's Falcon 9 rocket carrying the US Air Force X-37B space plane blasts off from Launch Complex 39A at Cape Canaveral.
Source: Picture Alliance / ZUMAPRESS.com / SpaceX

The historic Launch Complex 39A was the site of some of NASA's most historic launches. Pictured here is the space shuttle Endeavour at pad 39A in 2010.
Source: Image Credit: NASA / Bill Ingalls

1973: Skylab 4 on top of a Saturn 1B space vehicle launches from Launch Complex 39. Three astronauts in an Apollo command and service module are on board.
Source: NASA / Wallops

Today Port Canaveral remains a place of both commercial tourism and space exploration as SpaceX boosters and recovery ships dock at the port.
Source: Picture Alliance / Newscom / Stan Szeto

The space shuttle's external propellant tank being towed out of the Pegasus barge docked right across from the main assembly building.
Source: NASA

Connected by rail to the mainland, NASA's railroad train transports the last space shuttle solid rocket booster segments over the Indian River that forms the western water barrier of the Cape.
Photography: NASA / Kim Shiflett

Solid rocket booster recovery ship Liberty Star tows a solid booster of the space shuttle Endeavour past a docking Disney cruise ship.
Source: Picture Alliance / dpa / Bruce_Weaver

1965: A small trailer was used as the Visitor Information Centre
while permanent buildings were being constructed.
Source: NASA

1967: With a steady increase in demand
fuelled by the successes of Mercury, Gemini,
and Apollo, the flow of visitors expanded.
Source: NASA

1970: Apollo 17 commander Gene Cernan
waves to tourists in tour buses giving way to
his rover through the testing ground.
Source: NASA

2021: The Kennedy Space Centre adjacent to Cape Canaveral Air Force Station opens its door for people of all ages to enjoy the sci-fi theme park in the shadow of the space shuttle booster.
Source: Calvin L. Leake

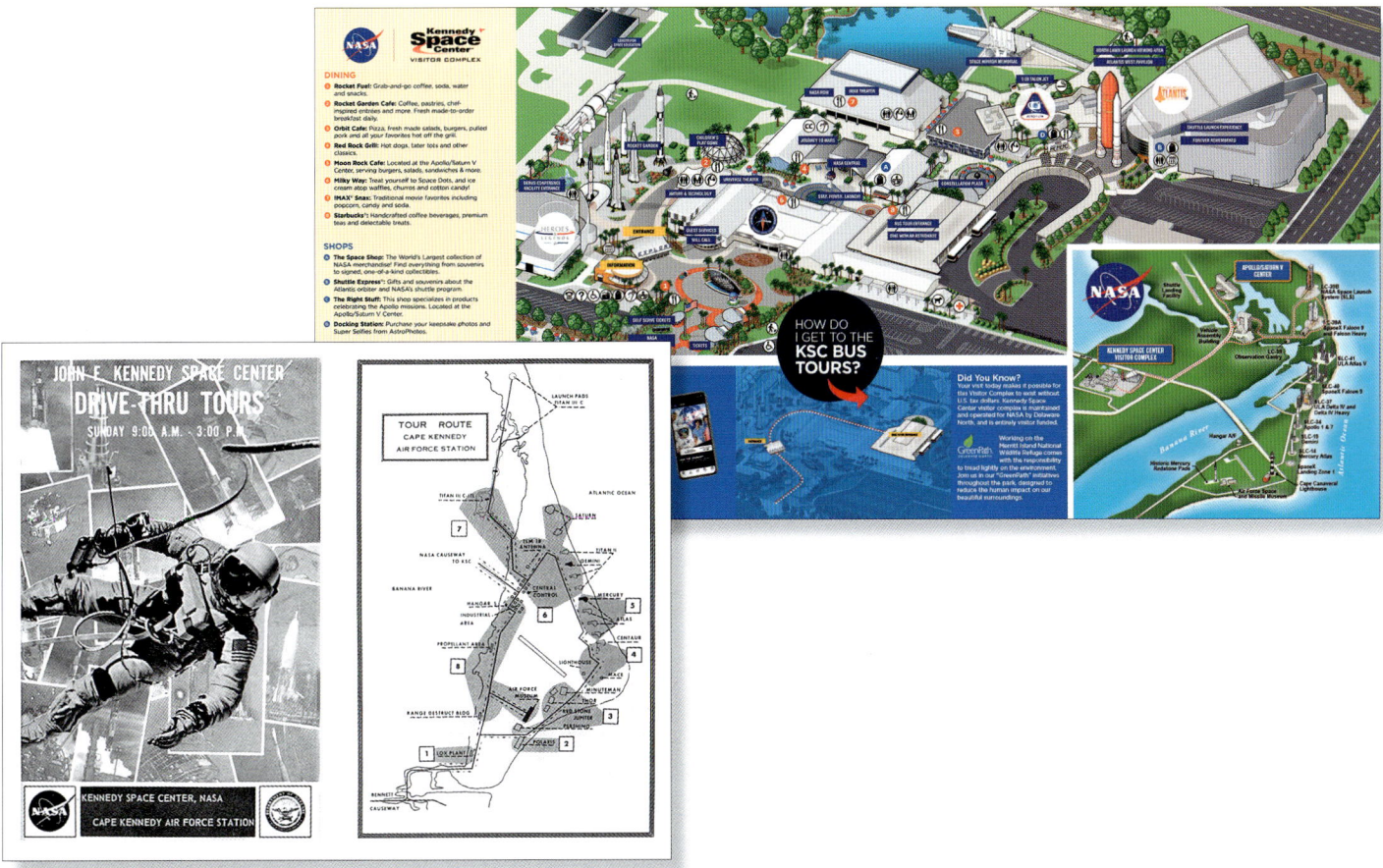

Few military facilities come with their very own visitor maps. Cape Canaveral's space tourism dates back to the 1960s.
Source: NASA

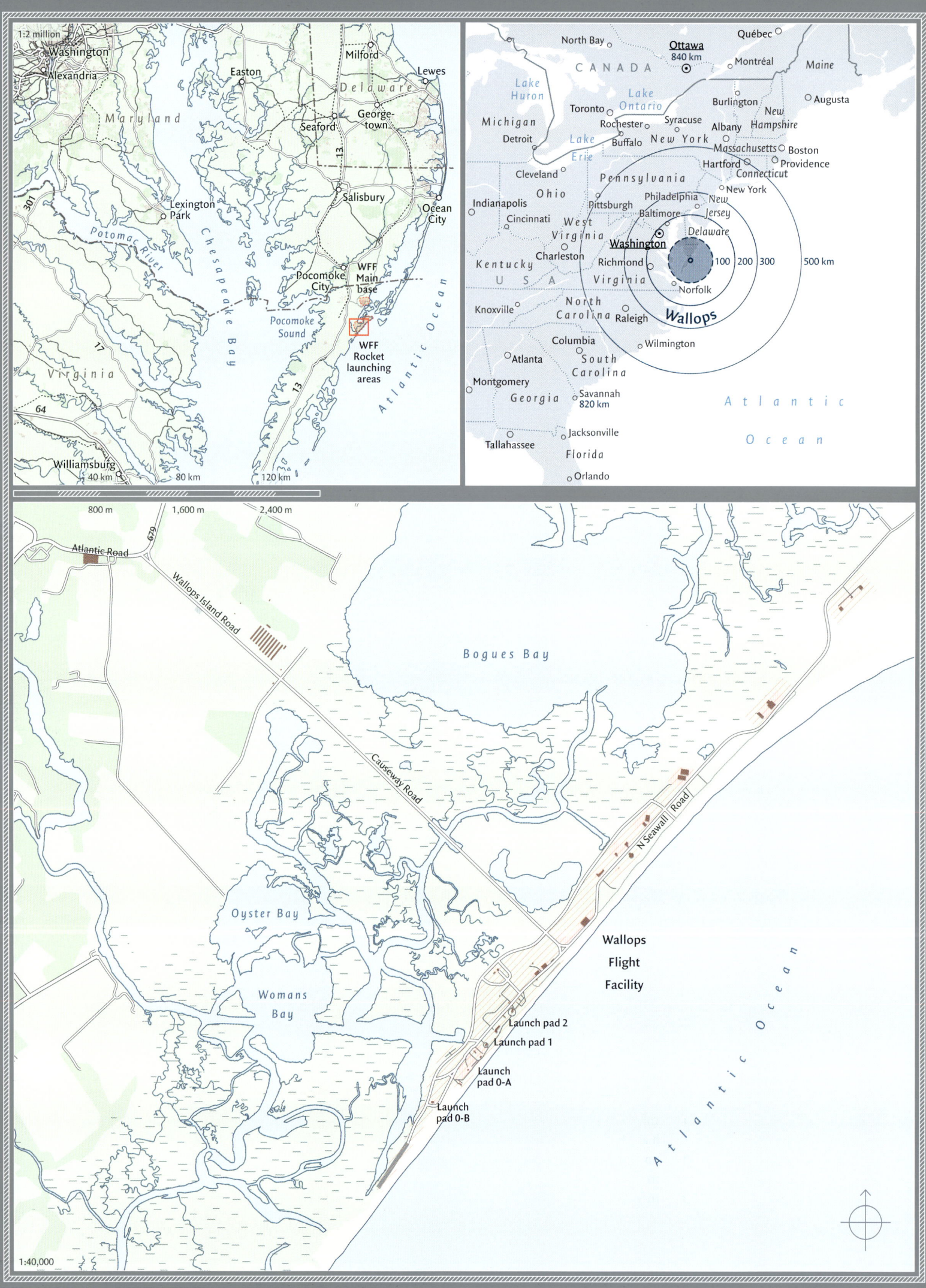

1:2 million

Washington
Alexandria
Easton
Milford
Lewes
Maryland
Delaware
Seaford
George-town
Lexington Park
Salisbury
Ocean City
Potomac River
301
Chesapeake Bay
Pocomoke City
WFF Main base
Pocomoke Sound
WFF Rocket launching areas
Virginia
17
64
Williamsburg
40 km   80 km   120 km
Atlantic Ocean
13

North Bay
Ottawa 840 km
Québec
CANADA
Maine
Lake Huron
Montréal
Michigan
Lake Ontario
Toronto
Rochester
Syracuse
Augusta
Detroit
Buffalo
Albany
New Hampshire
Lake Erie
New York
Massachusetts
Boston
Cleveland
Ohio
Pennsylvania
Hartford
Providence
Connecticut
Indianapolis
Pittsburgh
Philadelphia
New York
Cincinnati
West Virginia
Baltimore
New Jersey
Kentucky
Charleston
Washington
Delaware
USA
Richmond
100 200 300   500 km
Knoxville
Virginia
Norfolk
Wallops
North Carolina
Raleigh
Columbia
Wilmington
Atlanta
South Carolina
Atlantic
Montgomery
Georgia
Savannah 820 km
Ocean
Tallahassee
Jacksonville
Florida
Orlando

800 m   1,600 m   2,400 m
Atlantic Road
679
Wallops Island Road
Bogues Bay
Causeway Road
Oyster Bay
N Seawall Road
Wallops Flight Facility
Womans Bay
Atlantic Ocean
Launch pad 2
Launch pad 1
Launch pad 0-A
Launch pad 0-B
1:40,000

ATLANTIC SHORESIDE SITE

Name
Wallops Flight Facility

Location
Virginia, United States

Owner/Operator
National Aeronautics and
Space Administration

Elevation
2 m

Coordinates
37.8°N, 75.5°W

Time zone
GMT -4

Launches
116 satellites

Completion
1961

Wallops Island is the least well-known of the launch sites on the American mainland, but one with an illustrious history going back to 1672 when John Wallop, after whom the site was named, was deputy surveyor of Virginia. It started out in aviation as Chincoteague Naval Air Station and was taken over by NASA's predecessor, the National Advisory Committee for Aeronautics (NACA), on 27 June 1945. Its name and its parts have been through many evolutions (Station, Flight Centre, Flight Facility, Spaceflight Centre). The first of a prodigious total of 16,000 sounding rocket launches took place on 4 July 1945. Its features were similar to Cape Canaveral (qv) down the coast: a sand spit seafront with a bay and inland waterways, the Atlantic lapping its shores and launch pads right beside the ocean. Wallops Island is up in the far north-eastern corner of Virginia, bordering Maryland and Delaware and not far from New Jersey and

New York. The 2,506-hectare site has six launch areas with eight pads and 84 structures, assembly shops, payload processing areas, radar and optical tracking stations, fuel storage facilities, satellite tracking and data receiving stations, water towers, and blockhouses. There is a single integrated control centre, visitor centre, and launch view areas. The Scout rocket made Wallops. In 1958, NASA defined the need for a small, solid-fuel launcher to put its small scientific satellites into orbit. This became the Explorer programme. The four-stage Scout was first successfully launched from Wallops as Explorer 9 on 16 February 1961. Unlike normal American practice, it was assembled horizontally, as in the Soviet system. The Scout proved to be a reliable small satellite launcher (60 to 300 kg) and was not retired until 1994. During the early 1960s, Wallops hosted engineers from India and Pakistan. Wallops nearly succumbed to

Top:
What looks like a Mars rover is in fact a rocket atop a remote-controlled carrier rig. Despite its relatively large scale, it does not have to be carried by rail as required by its larger counterparts.
Source: NASA

Right:
A medium-capacity Antares rocket rigged for launch flanked by four light and sensor towers.
Source: NASA

congressional cuts in the 1980s, but orbital missions ceased towards the end of the Scout's life. Its obituary as a launch centre proved premature, for Wallops Island gained a new lease of life in the twenty-first century with the arrival of commercial companies contracted by NASA to supply the International Space Station. One of them was the Virginia-based Orbital Sciences Corporation, now part of Northrop Grumman, which built the Ukrainian-designed, Russian-engined Antares rocket to launch the Cygnus supply craft. Operational once again, regular orbital launches resumed at Wallops in April 2013.

2017: Antares, Orbital Sciences' first contracted cargo delivery flight, being rolled out to
the launch pad on a gloomy afternoon.
Source: NASA / Bill Ingalls

The relatively small airport facilities
on Wallops Island.
Source: NASA

Looking up the slim beach of the launch range,
one can see the North Atlantic up to Tom's Cove.
Source: NASA / Wallops

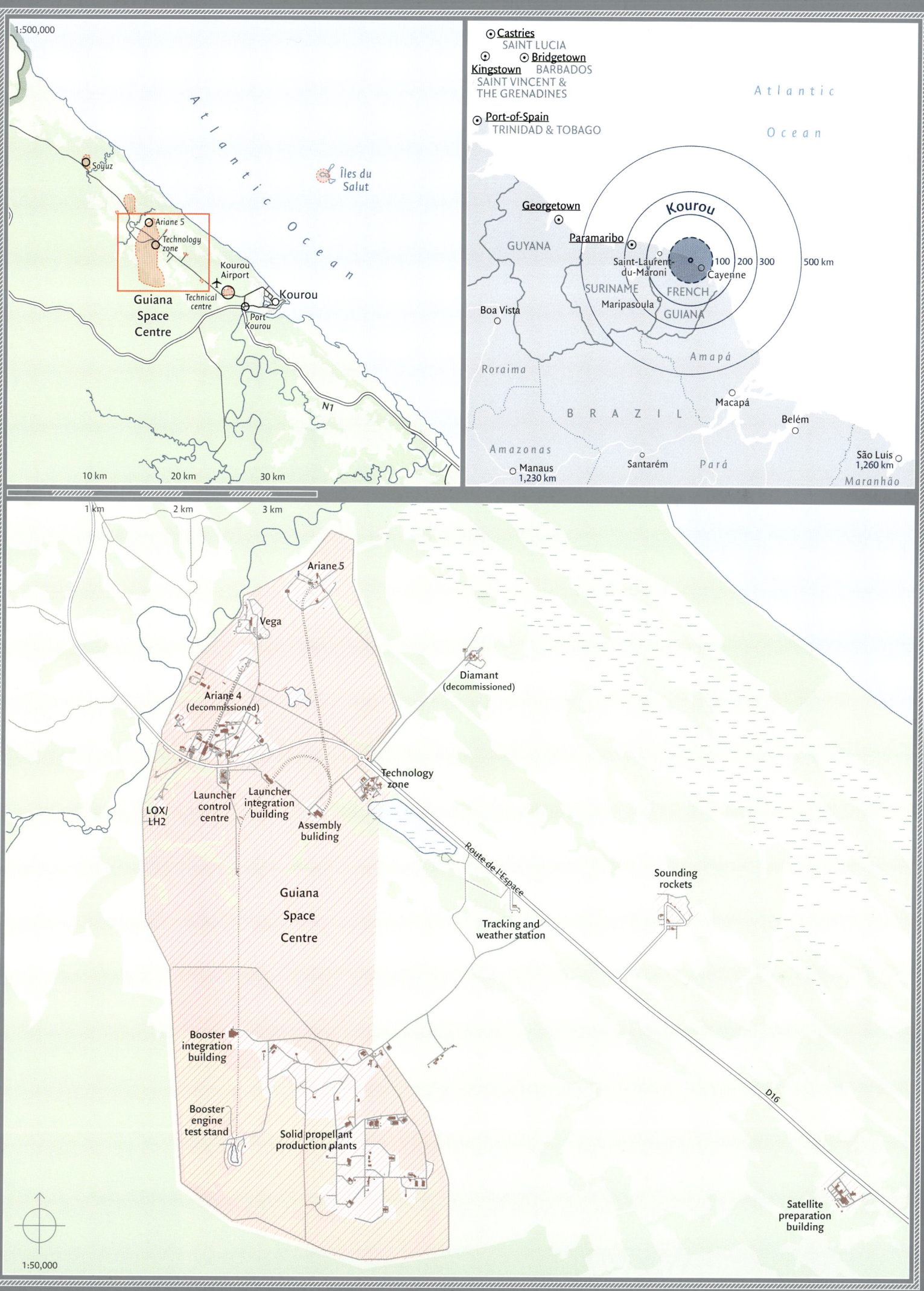

1:500,000

Atlantic Ocean

Soyuz

Ariane 5

Technology zone

Îles du Salut

Guiana Space Centre

Kourou Airport

Technical centre

Port Kourou

Kourou

N1

10 km   20 km   30 km

⊙ Castries
SAINT LUCIA
⊙ Bridgetown
BARBADOS
Kingstown
SAINT VINCENT &
THE GRENADINES

⊙ Port-of-Spain
TRINIDAD & TOBAGO

Atlantic
Ocean

Kourou

Georgetown
⊙

Paramaribo
⊙

Saint-Laurent-
du-Maroni
Cayenne

100 200 300   500 km

GUYANA

SURINAME   FRENCH
Maripasoula   GUIANA
Boa Vista

Amapá

Roraima

B R A Z I L   Macapá

Belém

Amazonas

Manaus   Santarém   Pará
1,230 km   São Luís
                                      1,260 km
              Maranhão

1 km   2 km   3 km

Ariane 5

Vega

Diamant
(decommissioned)

Ariane 4
(decommissioned)

Launcher
control
centre

Launcher
integration
building

Technology
zone

LOX/
LH2

Assembly
buliding

Route de l'Espace

Sounding
rockets

Guiana
Space
Centre

Tracking and
weather station

Booster
integration
building

D16

Booster
engine
test stand

Solid propellant
production plants

Satellite
preparation
building

1:50,000

1960: Wallops Island launch area number 3, used for Scout rocket firings.
Source: NASA / James Schulz

Left:
1960: Little Joe rocket with a Mercury capsule at its tip. The small rockets were used to test the escape systems.
Source: NASA

Right:
1970: Some 32 sounding rockets were launched at Wallops to conduct experiments during a full solar eclipse event.
Source: NASA

Kourou, French Guiana

EUROPE'S LAUNCH SITE IN THE SOUTH AMERICAN JUNGLE

Name
Guiana Space Centre

Location
French Guiana,
South America

Owner / Operator
European Space Agency
Centre National
d'Études Spatiales

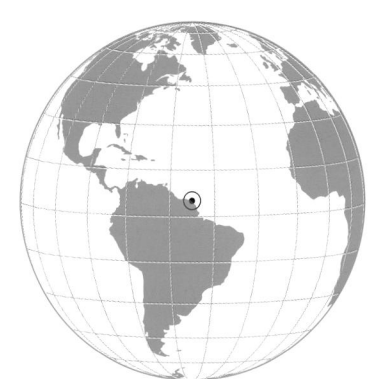

Coordinates
5.1ºN, 52.5ºW

Time zone
GMT-3

Launches
650 satellites

Completion
1970

As soon as the Évian Accords were signed and France gave up its colonial launch base in Hammaguir, Algeria (qv), it knew that it would have to find a new launch site. Sixteen were considered, from the Atlantic coast of France to French Polynesia. The winner was decided in 1964: Kourou in French Guiana, a colony in South America since 1664. The advantages of being so close to the equator (550 km) and launching out to sea outweighed the disadvantages of its steamy humidity, distance, and the need for port development at the capital city and port, Cayenne, to take arriving rockets. President de Gaulle visited and declared it a great French project.
The site allocated was 1,000 km², or one per cent of the Guiana land area – a windless open plain on a strip of coastal land 18 km northwest of Kourou on the road to Sinnamary. The strip is 29 km wide and 60 km long. Temperatures average 26ºC and rainfall is 2.9 m

per year. Inland is jungle, while off the coast are the notorious island penal colonies. Clearing and construction began in 1965, with the first launch, a Véronique sounding rocket, on 9 April 1968. The site had tracking stations in the small hills of Montagne des Pères 10 km away and Montabo, Cayenne, 60 km downrange. The Kourou end had the technical centre and Jupiter control room. The first orbital launch from Kourou was on 10 March 1970 when the Diamant rocket, hitherto launched from Hammaguir, put two small German satellites into orbit. Diamant made six successful orbital flights from there, the last in 1975.
When Britain pulled out of the European Launcher Development Organisation (ELDO) in 1968, it was decided to move launches of the large European rocket, the Europa, from Woomera, Australia (qv), to Kourou. The first orbital attempt was made from there

Christmas 1979: Surrounded by jungle rather than pine trees, engineers watch the first successful launch of Europe's launcher, Ariane 1, from Kourou.
Source: ESA

4 June 1996: Only 37 seconds after launching, ESA's first Ariane 5 rocket self-destructed due to a malfunction in the control software. Not far off, the launch site inhabitants of a village watch the fall-out of the rocket spread across the sky.
Source: Picture Alliance / dpa / epa / Antoine_Cercueil

Right:
2019: Construction of the 'next generation' launch pad. The new Ariane 6 rocket and assembly have been optimised to allow for more and cheaper launches.
Photography: Picture Alliance / abaca / Aventurier Patrick / ABACA

Next double page:
French Guiana is a territory of France, thus it is the French Army that secures the launch pad of an Ariane launch.
Source: Picture Alliance / abaca / Aventurier Patrick / ABACA

on 5 November 1971, unsuccessfully and the programme was cancelled two years later.

Kourou now changed from being a French project to a European one. Legally, it is now a facility of the French space agency, Centre National d'Études Spatiales (CNES), made available to the European Space Agency (ESA). When the new European Space Agency (ESA) was established in 1975, it inherited a proposal for a new, French-led rocket, the L3S, later called Ariane, designed to launch communications satellites into 24-hour orbit, for which the equatorial launch site in Kourou was ideal, as well as other European payloads. Kourou came back to life in time for the arrival of the first Ariane by barge from Le Havre. Ariane was encased in a new mobile launch tower. Its home was called Ensemble de Lancement Ariane (ELA) 1. The first Ariane launched in the warm, equatorial night sky on 24 December 1979. Kourou has never looked back and has been constantly busy as Europe's spaceport ever since, with Ariane 1 followed by the 2, 3, and 4 versions in 1984, 1986, and 1988, respectively. The high level of activity required the construction of a second launch pad, ELA-2, inaugurated in 1986 and intended to launch up to eight rockets a year. ELA-2 comprised an 80-metre-tall assembly hall, a double railway track to bring the rocket on an hour-long journey to the pad, turntable, and a 74-metre-tall handling platform and umbilical tower. ELA-2 was taken out of service with the last Ariane 4 in 2003.

The European Space Agency then moved on to a much larger rocket, quite different in design from the 1-4 series: the Ariane 5, which was able to launch two six-tonne communications satellites at a time, or quite large scientific payloads of up to 12 tonnes.

Ariane 5 was also intended to fly the manned European spaceplane, Hermes, from Kourou, though this project was abandoned. In contrast to the Ariane 1-4 series, which used storable nitric fuels, the Ariane 5 used more powerful hydrogen fuel for its first stage and was assisted by two large solid rocket boosters.

This monster rocket required a third launch pad, ELA-3, started in 1988 and completed 1992. This was a massive construction project, requiring: 2,100 ha of land, 40 km of new roads, a 7 km railway, 4m m³ of earthworks, 80,000 m³ of concrete, 7,000 tonnes of metal, the canalisation of the Kourou River to take the barges with the centre stage, an artificial lake, and a new hydroelectric power plant. First to be built was the propellant plant, which covered an area of 300 ha of buildings, mixers, casting pits, and control facilities to pour the sludge-like fuel for the solid rocket boosters, which was made on site. The biggest building was the 58-metre-tall Launcher Integration Building, where the solid rocket boosters were attached to the first and second stage. The assembled rocket was moved to the 83-metre-tall Final Assembly Building, where its payload was attached in air-conditioned cleanroom conditions. The Ariane 5 was then brought 2,800 m to the pad in a crawler on a wide-gauge railway. It was fuelled from tanks below the mobile launch table – the Russian system developed for Zenit. The final part of the new facilities was a launch control centre called Jupiter 2, 15 km away and with a two-tier control room, an inner small control room of four ranks for consoles, and, behind a glass screen, large auditorium seating for journalists and visitors. A giant processing facility was added in 2001, blandly called S5, with a floor area of 3,800 m² to handle large cargoes destined for the International Space Station.

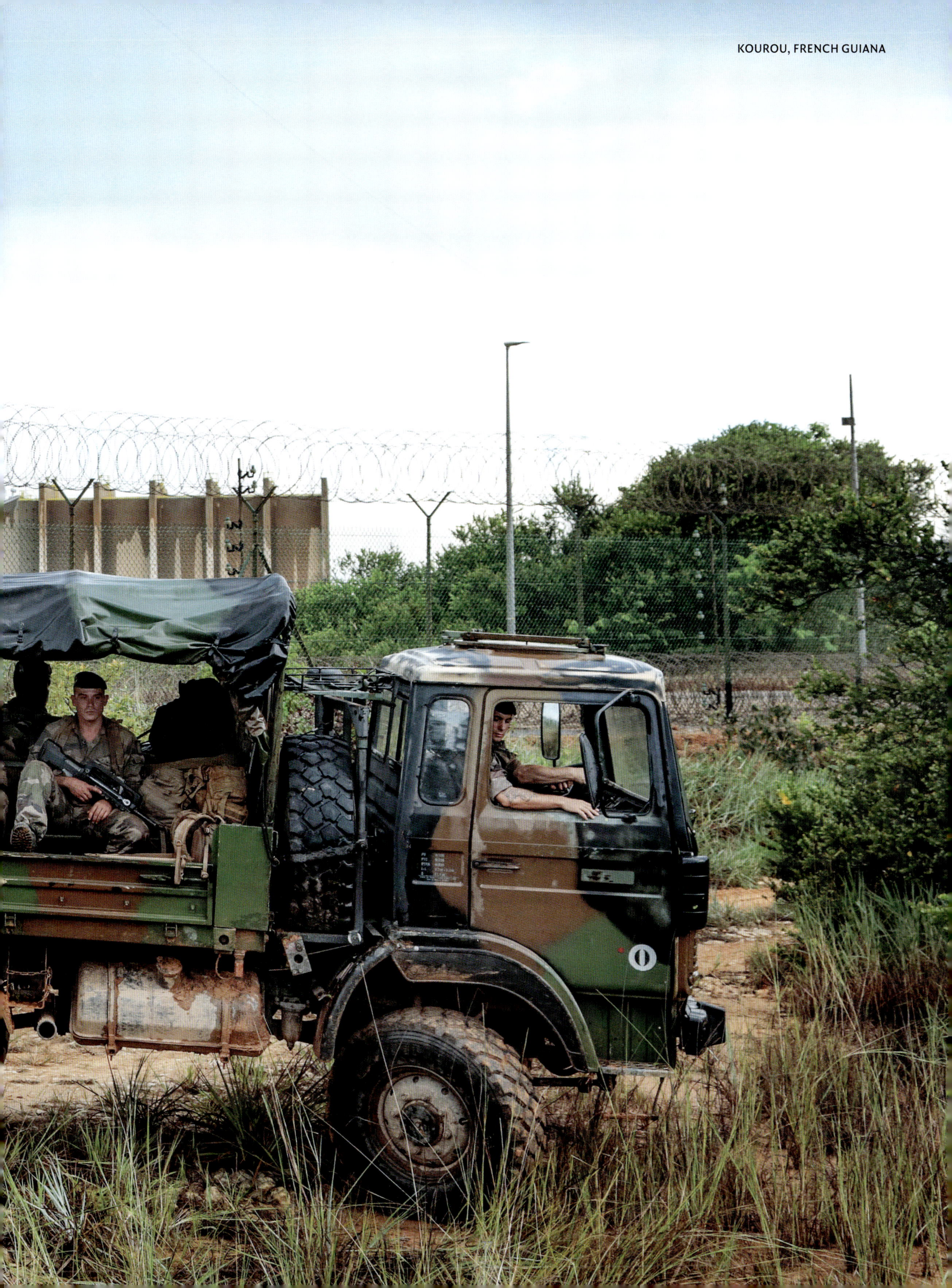

Ariane 5 first flew on 4 June 1996, exploding catastrophically, but it quickly became a reliable and successful launcher with over one hundred launches to date. The next stage of Kourou's development was improbable, even unimaginable. Ariane 5 left Europe without a medium-lift rocket, so French companies developed a relationship with the Russian Progress company in Samara to fly medium class European payloads on the R-7 Soyuz rocket. Flying the Soyuz from Kourou was a logical extension of this experience, approved by the European Space Agency in 2003. The project faced hurdles in terms of financing, security, and the management of a European and Russian workforce. The Soyuz system was so distinctive as to require a new, 120 ha launch site 12 km north of ELA-3, with clearing starting in 2005. The project required 20,000 m² of floorspace, 96,000 m² of roads, and the excavation of 200,000 cm³ of rock for the 26-metre-deep flame trench. The last part was literally the toughest. It required the blasting of granite for the huge flame pit required by the Soyuz, often in tropical rain. It was one of the biggest construction projects in the world at the time. Called ELS (Ensemble de Lancement Soyuz), it comprised elements familiar to existing Soyuz sites: a vehicle assembly building (MIK), fuel storage tanks, a liquid oxygen production plant, technical rooms, a 600-metre railway line to the pad, and a tulipan launch site with traditional arms. The principal difference with other Soyuz pads was a 42-tonne, 56.6-metre-high mobile tower for the vertical installation of the Fregat upper stage and payload – necessary to protect them from humidity and rain. It was a big logistical operation, with Soyuz brought there from Samara, the Fregat from Moscow, and payloads from Le Havre, France. The project came to fruition on 21 October 2011 when the world saw a Soyuz rocket rise above the tropical forest to put into orbit a prototype Galileo navigation satellite. A total of 22 launches were accomplished by 2020. Soyuz meant that Europe's Guiana-based launcher fleet now had both a medium-lift (Soyuz) and heavy-lift (Ariane 5) launcher. For many years, Italy, an expert in solid-fuel rockets, pressed for Europe to have a light-lift launcher, the Vega. Approved in 1998, Vega eventually flew on 13 February 2012 and made 14 successful launches before its first failure in 2019. Vega used the original launch site at Kourou, ELA-1, retired 1989, renovated, and renamed Ensemble de Lancement VEGA (ELV).

In 2012, the ESA approved a new heavy-lift rocket, the Ariane 6, intended to be a commercially-competitive launcher able to launch payloads in two versions in the four to 16 tonne range. This required a new launch site: ELA-4. Construction began in 2017, with a view to completion in time for the first launch in 2020. A new 170 ha site was cleared 17 km from Kourou and 4,000 m from the Ariane 5 pad. The rocket reverted to the Soviet system of horizontal integration in a Launch Vehicle Assembly Building 112 m long, 20 m high, and 41 m wide before transfer to an enormous 8,200-tonne mobile gantry. The pad is 28 m deep, 200 m wide, and has a 700-tonne launch table with four lightning masts at the side. Kourou now has four operational pads: ELV (Vega; formerly ELA-1), ELS (Soyuz), ELA-3 (Ariane 5), and ELA-4 (Ariane 6), with ELA-2 retired. It is possible to visit Kourou, which has a space museum, and arrange advance-booking tours of the launch sites and facilities, but given what it can offer, it is surprisingly undeveloped as a tourist destination.

1:60,000

Alcântara
Space Centre
(Centro de Lançamento
de Alcântara)

Launch
Pad A1

1,200 m   2,400 m   3,600 m

Cayenne
1,190 km
FRENCH
GUIANA
Atlantic Ocean
Amapá
Macapá
Alcântara
Belém
100 200 300   500 km
São Luís
Fortaleza
680 km
Pará
Caxias
Marabá
Teresina
Ceará
Rio Grande
do Norte
Maranhão
Paraíba
B R A Z I L
Piauí
Pernambuco
Tocantins
Petrolina
900 km
Maceió
Mato
Grosso
Palmas
Aracaju
Bahia
Sergipe

10 km   20 km   30 km
Cedral
MA-304
Mirinzal
308
MA-305
Guimarães
Baía do Cumã
A t l a n t i c
O c e a n
Rio Pericumã
Alcântara
Space
Centre
Launch
pad A1
Golfão
Maranhense
MA-106
Rio Itapetininga
Rio Salgado
Alcântara Space
Centre Airport
Alcântara
MA-211
Cajual
Island
Raposa
MA-203
Bequimão
São Luís
Paço do
Lumiar
MA-106
MA-201
São José
de Ribamar
M a r a n h ã o
Marechal Cunha
Machado Int. Airport
135
Baía de
São Marcos
S ã o L u í s I s l a n d
Baía de
São José

1:500,000

BRAZIL'S EQUATORIAL LAUNCH SITE

Name	**Coordinates**
Alcântara Space Centre	2.3°S, 44.4°W
Location	**Time zone**
Brazil, South America	GMT-3
Owner / Operator	**Launches**
Agência Espacial Brasileira	None successful
Elevation	**Completion**
32 m	1982

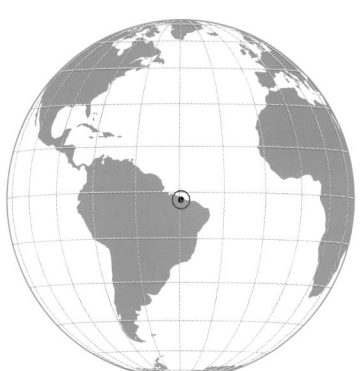

Alcântara, Brazil, is the only indigenous launch site in the South American continent. Kourou (qv), up the coast, is larger, but a former French colony. Alcântara is probably the launch site with the most unhappy history.

Brazil's space programme can be traced to the 1960s and it is the South American country with the largest investment in space research. A small sounding rocket site was developed at Barreira do Inferno in Natal (5.9°S, 35.2°W) on the easternmost tip of the country from 1965 to 1989, but it was close to urban areas. A space plan was adopted in 1979 for the development of a launcher (Veículo Lançador de Satélites; VLS), satellite, and a larger, more suitable launch centre. This became Alcântara Space Centre, set on the Alcântara peninsular on the Atlantic coast in the state of Maranhão. It had the advantage of being the closest to the equator in the world, only 220 km away, as well as ease of access and low population density. The site covers a huge area measuring 620 km², the main part having four pads (sounding rocket, VLS, and two others), a 33-metre-tall, 380-tonne, 10-storey mobile integration tower on rails, and a launch tower, with facilities for payload preparation, engines, and fuelling. There is a control centre, ground satellite station, and a 2.6-km runway for an airfield. It is wet from April to May, but dry the rest of the year. The first sounding rocket was fired from here in 1990, its year of opening. The first attempt to launch a satellite was made on the four-stage solid-fuel VLS-1 in November 1997 and a second was made in December 1999; both were unsuccessful. Preparations were underway for a third attempt on 22 August 2003 when the fourth stage fired on the pad, causing the whole vehicle to explode and killing 21 engineers and destroying the mobile tower.

Smoke overshadows the Brazilian jungle: Murphy's
law haunts both Brazil's space ambitions as well
as its so-called 'cursed spaceport'. The image on
the right shows the tragic aftermath of the 2003
rocket explosion.
Source: Picture Alliance (Top), ESA (right two images)

Left:
Aerial view of the launch tower at the base of
Alcântara in the coastal jungle of northern Brazil.
Source: Picture Alliance / AP Images / Eraldo Peres

Next double page:
Brazilian Air Force airmen in front of the launch tower.
Source: Picture Alliance / AP Images / Eraldo Peres

The VLS programme was subsequently prohibited by the United
States for many years. Attempts were made to revive Alcântara as
a launch site with other countries as partners, the most promising being the Ukrainian Tsyklon 4 rocket. A pad was even built,
but the project collapsed at a late stage. The large mobile tower
was reconstructed, and the site is still used for sounding rockets.
Alcântara is a fully-equipped launch centre, which, coupled with
its location, means that it is unlikely to stay idle forever.

BALTO-MEDITERRANEAN

Baikonur

Plesetsk

Kapustin Yar

Caspian Sea

Sem

Baltic Sea

Black Sea

ISRAEL

Palmachim

Peenemünde

Mediterranean Sea

GERMANY

ALGERIA

Hammaguir

Sriharikota
⊙

Kulasekharapatnam
⊙
⊙
Thumba

Persian
Gulf

Indian

Ocean

Equator

Red Sea

⊙
San Marco

Atlantic

Ocean

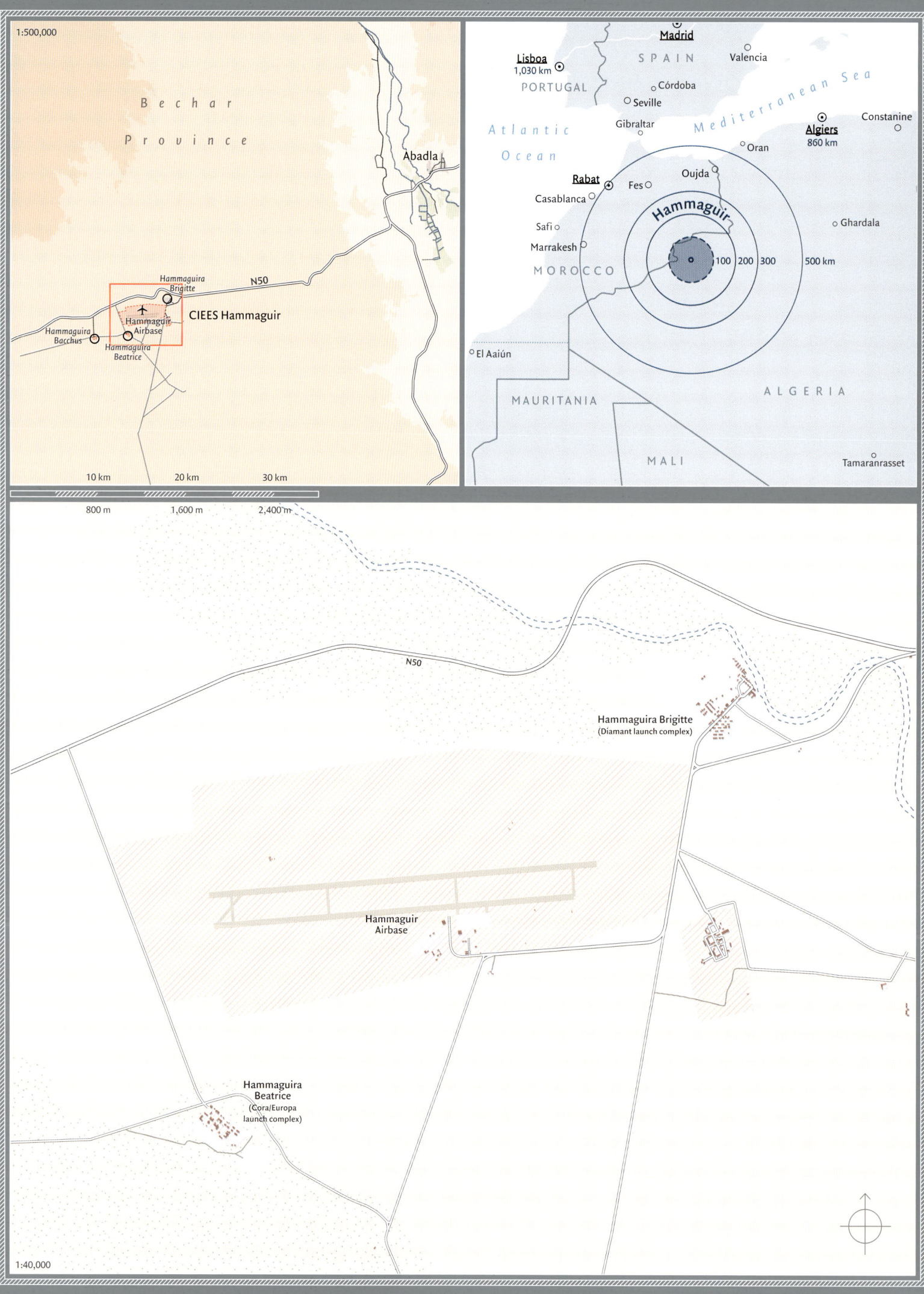

1:500,000

Bechar

Province

Abadla

Hammaguira
Brigitte

N50

CIEES Hammaguir

Hammaguir
Airbase

Hammaguira
Bacchus

Hammaguira
Beatrice

10 km   20 km   30 km

Madrid

Lisboa
1,030 km

SPAIN

Valencia

PORTUGAL

Córdoba

Seville

Mediterranean Sea

Atlantic
Ocean

Gibraltar

Constanine

Algiers
860 km

Oran

Rabat

Fes

Oujda

Casablanca

Hammaguir

Ghardala

Safi

Marrakesh

100 200 300   500 km

MOROCCO

El Aaiún

ALGERIA

MAURITANIA

MALI

Tamaranrasset

800 m   1,600 m   2,400 m

N50

Hammaguira Brigitte
(Diamant launch complex)

Hammaguir
Airbase

Hammaguira
Beatrice
(Cora/Europa
launch complex)

1:40,000

Abadla

FRANCE'S DESERT SITE IN NORTH AFRICA

Name
Centre Interarmées d'Essais
d'Engins Spéciaux

Location
Algeria, North Africa

Owner / Operator
French Air Force

Elevation
744 m

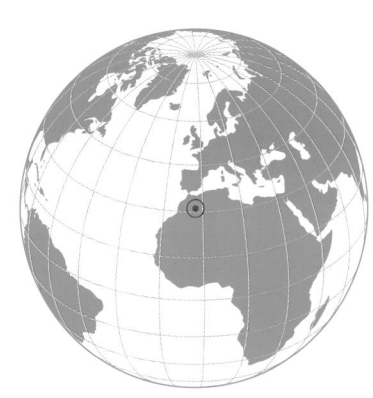

Coordinates
31°N, 8°W

Time zone
UTC+1

Launches
4 satellites

Completion
1965–1967

Most people have no trouble identifying the USSR and the United States as the first countries from which satellites were launched. Few would identify Algeria as the third. After the Second World War, France began to build a missile and space programme, first establishing a missile test base in 1947 in Algeria, the largest country in Africa, much of which is desert. Algeria was then part of France, although ever more engulfed in what became a violent struggle for its independence. France chose an area of low population density, establishing launch sites in the hot, dry desert plateau at the western end of the Grand Western Erg: Colomb-Béchar (two pads) and 120 km away in Hammaguir (Hammada du Guir in Arabic; five pads). Between the two, 231 rockets and missiles were fired. Hammaguir comprised a village, an airport to ferry in personnel, launch pads, storage areas, housing for 600 people, and recreational facilities. Sounding rockets, the best-known being the Véronique, were launched from 1952, reaching 70 km altitude. During the International Geophysical Year (IGY), 15 Véronique AGI (Année Géophysique Internationale) were fired from Hammaguir to an altitude of 210 km. From 1963, biological experiments were carried out, launching Félicitte, the first space cat, and Hector the rat. Most famously, Hammaguir was the launch site for the third country (France) to put its own satellite into orbit. On 26 November 1965, the three-stage Diamant A rocket launched the Astérix 1 satellite. Hammaguir was also home to France's next three satellites: Diapson (February 1966), Diadème 1, and Diadème 2 (February 1967) – all on the Diamant. Satellite launches took place beside a lighthouse-shaped cement blockhouse, with launch attachments scattering like shards on take-off. Ground tracks headed over the Mediterranean toward a tracking station in Beirut (first two launches) and over Niger (second two).

One of the interferometers at Hammaguir presents a striking image against the flat landscape of the desert. The installation was used to determine the position of rockets and satellites. When the National Centre for Space Studies left Hammaguir in 1967, it dismantled and packed up the interferometers and took them back to France.
Source: Encyclopedia of Space (1968), p. 79

The main control station or block house supporting launch operations at Hammaguir was in a location separate from the main living area.
Source: Encyclopedia of Space (1968), p. 80

Hammada du Guir ('the base of life') was the original plan for France's desert launch city for up to 600 people. However, it was decided in 1958 to spread out the facility into five individual shooting ranges.
Source: CNES

Under the independence settlement, the Évian Accords (1962), France kept Hammaguir on lease until 1 July 1967 when it was then required to evacuate. The French space agency, Centre National d'Études Spatiales (CNES), began searching for alternate sites as early as 1962, eventually settling on Kourou, French Guiana (qv), the subsequent site of French launches and from where the Diamant B was fired in 1970. Hammaguir is south of the road between Tindouf and Béchar. The installations appear to still be there. Lights show up on NASA's night maps, but it is not a destination for tourists or space archaeologists.

HAMMAGUIR COLOMB-BECHAR

Premier cosmodrome français

(De notre envoyé spécial Pierre LOUIS)

Cet insigne est commun à tous les personnels militaires appartenant au C.I.E.E.S. (Centre Interarmées d'Essais d'Engins Spéciaux).

— H moins une minute du tir !

Tout en surveillant attentivement le passage au vert des innombrables voyants lumineux du pupitre devant lequel il est installé, le véritable «chef d'orchestre» des tirs de fusées DIAMANT — capitaine Robert, 36 ans — égrène maintenant les dernières secondes du « compte à rebours » (ou, plus exactement : du décompte de la chronologie) :

— 5, 4, 3, 2, 1, zéro. Feu !

A 5 kilomètres du P.C.C.T. (Poste Central de Commandement de Tir) près duquel je suis installé, apparaît une fumée noirâtre qui, très vite, devient rousse. Trois secondes s'écoulent et, de cette fumée semble surgir une fusée qui s'élève tout droit dans le ciel sans nuages du Sahara, lentement d'abord, puis de plus en plus vite : c'est une fusée Diamant, porteuse de notre satellite « D 1 », qui vient de quitter la base de lancement « Brigitte » !

Encore un peu de patience et, en même temps que les civils et les militaires qui ont travaillé en équipe pendant des années en vue de cet essai, j'apprendrai son éclatante réussite : pour la première fois dans l'histoire de la conquête spatiale, une fusée française a placé sur orbite un satellite lui aussi français ! Dorénavant le champ de tir d'Hammaguir - Colomb-Béchar mérite bien le titre de « premier cosmodrome français » !

SUITE PAGE 29

C'est en plein désert, à 120 kilomètres au sud-ouest de Colomb-Béchar, qu'a été construite la base-vie du site d'Hammaguir. Sur cette photo aérienne, vous pouvez localiser :

1. La route goudronnée reliant la base-vie aux différentes installations du champ de tir «B2» et à Colomb-Béchar.
2. Le poste de contrôle, où sont vérifiées entrées et sorties.
3. Les bâtiments de la compagnie de la Légion

Etrangère qui assure la sécurité de la base-vie et du champ de tir.
4. Le stade.
5. L'aire d'atterrissage pour les hélicoptères.
6. La salle de cinéma en plein air.
7. Tennis.
8. Mess-hôtel pour ingénieurs, techniciens et officiers.
9. Piscine.

10. PC de la base-vie.
11. Bâtiments administratifs et logements divers.
12. Garage, ateliers et pompe à essence.
13. Chapelle.
14. Barraques Fillods pour logement de la troupe.
15. Parc à voitures.
16. Dépôt de carburants.
17. Château d'eau.

Hammaguir, Colomb-Béchar. Premier cosmodrome français article by Pierre Louis, with detailed information about the launch site in Algeria.

Source: Tintin, No. 917, 17 May 1966, p. 25.

France's first astronauts: Hector the rat (launched in 1961) and Félicette the cat (launched in 1963). The programme was the French response to Laika, the space dog launched by the Soviet Union and the first animal to orbit Earth. Unlike Laika's no-return mission, Félicette experienced a suborbital flight and was recovered 13 minutes after take-off.
Source: CNES

One of the space rats being inserted into a Diamant rocket nose cone.
Source: CNES

Scientists in front of a Cora rocket, which was the predecessor to the first rocket to be constructed by the European Launcher Development Organisation – the precursor of the ESA.
Source: CNES

Early French rocket launches conducted in the 1960s: Such rockets were the predecessors to the Diamant, which carried the first French satellite to space. A thin chain link fence marks the border of the Diamant launch complex from the surrounding sand. It was from here that France entered space as the third country to successfully launch a satellite.
Source: CNES

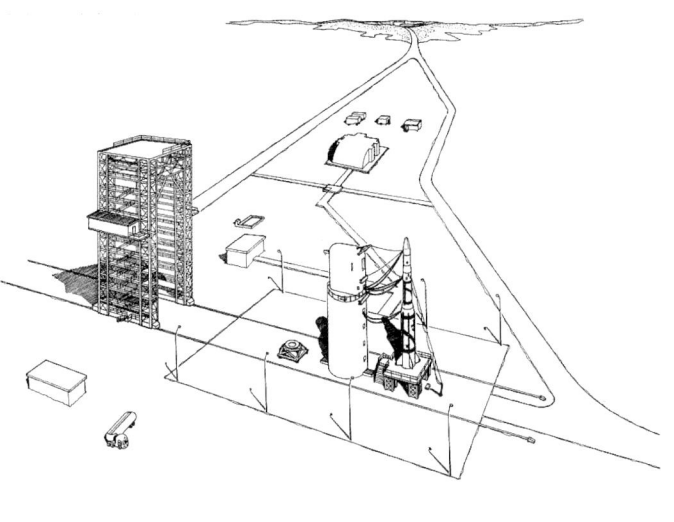

The Diamant A launcher on its launch pad. The rockets
Diamant A and B were the first satellite launchers to be
researched, constructed, and tested outside of the US
and the USSR, thus enabling France to become the third
global power in space exploration. Diamant was a three-
tier launcher with an equipment bay and a nose cone that
shielded the satellite. Its various upgraded versions led
to the designation of systems, anticipating the current
launcher, Ariane. The rocket Diamant A had been qualified
by the launch of the satellite Astérix-A1 on 26 November
1965 from the Hammaguir base. The last launch took place
in 1975 from the Guiana Space Centre.
Source: SEREB

The first launch of Diamant A on 26 November 1965,
Hammaguir. France became the third global power
in space exploration.
Source: ECPA

Cyclope – the large dish antenna for
receiving telemetry from rockets –
had a dish measuring 18 m in diameter.
Source: Musée de la Poste

A Diamant rocket on a French postal stamp.
Design proposal (left) and final print (right).
Designed by Claude Durrens, 1965.
Source: Musée de la Poste

After France conducted its first nuclear test in the southern desert of Algeria in 1960, it followed the two great nuclear powers of the Soviet Union and the USA into space. The French space programme was honoured with several stamps – a media of Cold War propaganda.
Source: CNES

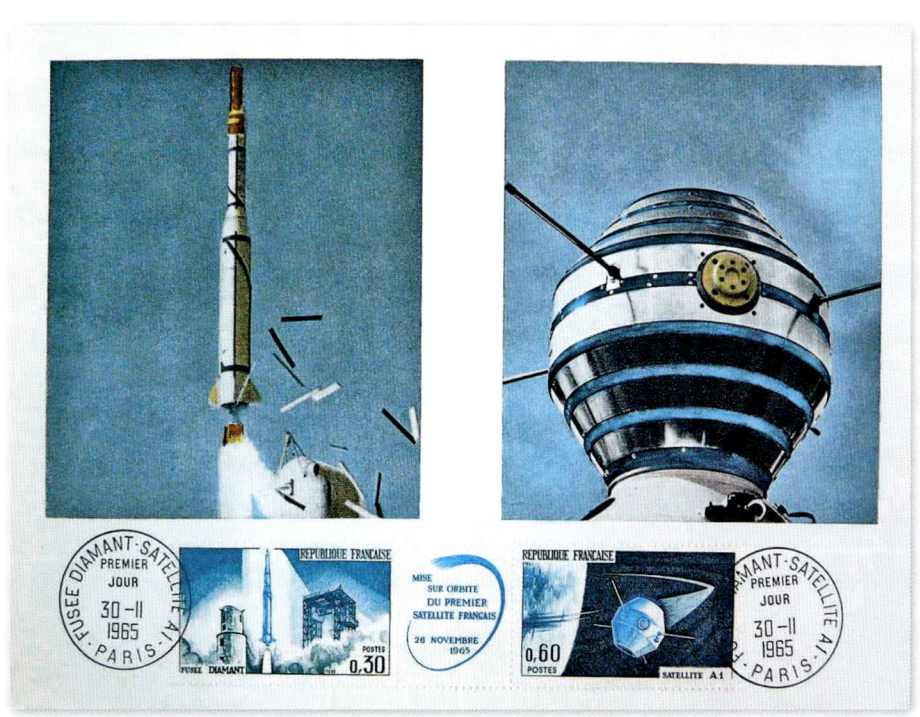

Top:

The Centre Interarmées d'Essais d'Engins Spéciaux (CIEES) was used for the launch of the four Diamant A launchers and was closed in 1967. Under the terms of the protocol signed on 29 November 1961 between the Minister of the Army and the Minister for Scientific Research, CNES began using the facilities of two military firing ranges: the Centre Interarmées d'Essais d'Engins Spéciaux (CIEES) and the Centre d'Essais et de Recherches d'Engins Spéciaux (CERES) on the Île du Levant. The CIEES, founded on 24 April 1947, operated under the Direction des Recherches et des Moyens d'Essais. It comprised two complexes: Colomb-Béchar and Hammaguir. The first one included a B1 firing range of modest dimensions and limited to low performance firings. The second, located 110 km from Colomb-Béchar, was installed on a vast plateau and had a base and a B2 firing range for all kinds of vehicles. During the period 1960–1961, more than 1,500 missiles were launched from these two bases.

Source: CNES / Louis Laidet

Right:

Topaze was the second test vehicle of the gemstone family. It was composed of a gunpowder monowheel, driven by four mobile nozzles that made it possible to study the problems relating to propulsion with powder. Topaze was also the first French guided rocket. Thus, it could be fired from a base, whereas the VE10 and VE110 (Agate) used a ramp. Two versions were successively produced: the VE111C (C = short) with an NA802 engine and the VE111L (L = long) with an NA803 engine. Six examples of the VE111C, produced between the end of 1962 and the end of 1963, were dedicated to the development of steering by turning the tubes. Ten VE111s had been planned for this, but after a series of six successes, it was decided to reserve the four remaining ones for the study of the piloting of more unstable machines such as the MSBS. These tests, carried out in 1964 with the VE111CI (CI = short unstable) version, had three successes out of four shots. The VE111L was successfully tested twice in 1963–1964. Two other examples were used to test a complete inertial guidance system. These two firings in the VE111LG (LG = long guidance) version were also successful in May 1965. The 14 firings took place at Hammaguir in Algeria at the Centre Interarmées d'Essais d'Engins Spéciaux (CIEES).

Source: CEL / CNES

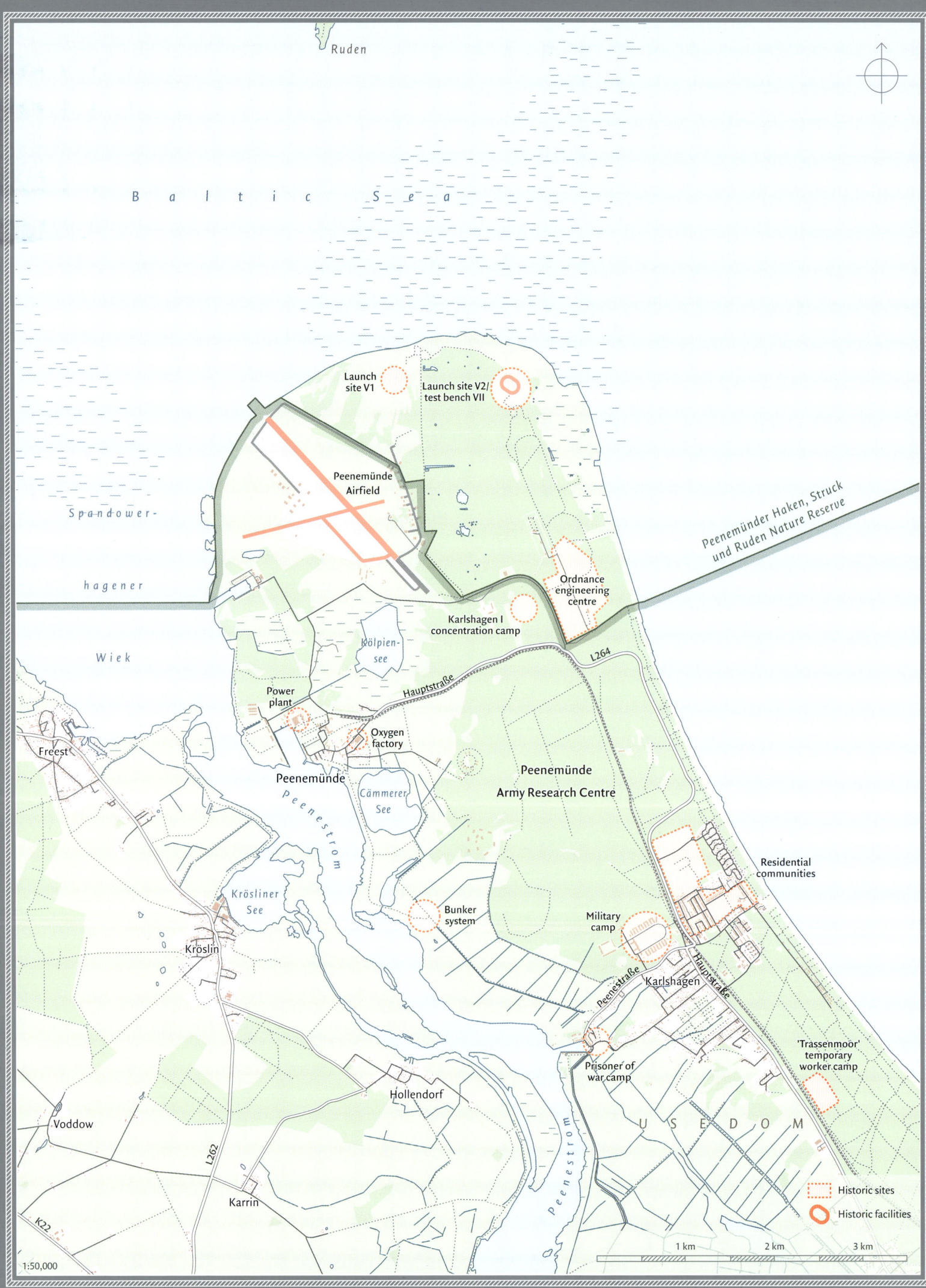

Ruden

B a l t i c   S e a

Spandower-

hagener

Wiek

Launch
site V1

Launch site V2/
test bench VII

Peenemünde
Airfield

Peenemünder Haken, Struck
und Ruden Nature Reserve

Ordnance
engineering
centre

Kölpien-
see

Karlshagen I
concentration camp

L264

Power
plant

Hauptstraße

Freest

Oxygen
factory

Peenemünde

Cämmerer
See

Peenemünde
Army Research Centre

Peenestrom

Residential
communities

Krösliner
See

Bunker
system

Military
camp

Kröslin

Peenestraße

Hauptstraße

Karlshagen

'Trassenmoor'
temporary
worker camp

Hollendorf

Prisoner of
war camp

U S E D O M

Voddow

L262

Peenestrom

Karrin

K22

Historic sites

Historic facilities

1:50,000

1 km   2 km   3 km

BALTIC HOME OF THE MODERN ROCKET

Name Heeresversuchsanstalt Peenemünde		**Elevation** 3 m
Location Germany, Europe		**Coordinates** 58.1°N, 13.8°E
		Time zone GMT+2
Owner/Operator Heereswaffenamt, Nazi Germany		**Completion** 1936

It was from Peenemünde that the first modern rocket was fired on 3 October 1942 – by remarkable coincidence 25 years and one day before Sputnik. This was the German wartime A-4, better known and feared as the V-2. That day, the A-4 took off, bent over in its climb, reached 85 km altitude – the boundary of space – and crashed into the Baltic Sea 190 km away and 296 seconds later. 'Today, the spaceship was born,' was said by those who watched the spectacle. The engineer responsible for the A-4 was Wernher von Braun. Knowing Germany's north coast well, he also personally selected the launch site. Moving from the Wehrmacht testing grounds in Kummersdorf near Berlin to more distant, safer locations, Von Braun first fired small rockets from the island of Borkum in the North Sea and then Greifswalder Oie in the Baltic. For larger rockets, a real rocket base was needed. Peenemünde ('the town

at the mouth of the Peene River') was a tongue of land of pine-woods and duck-populated reedy promontories stretching out on the island of Usedom into the Baltic Sea, 97 km north-west of the town of Stettin (German) or Szczecin (Polish), the local town being Karlshagen. In 1935, construction began of engine test stands, launch pads, observation posts, sheds, factories, workshops, a wind tunnel, a flight mechanics computation office, a power station, and an airfield. There was also a substantial housing area. The first two attempts to launch the A-4 in autumn 1942 failed. The 3 October launch was the third test and used *Prüfstand VII* ('test stand VII') in a forest clearing. It took almost two years of further firings to ready the A-4 for military use. The Royal Air Force flew over Peenemünde and, suspecting its true purpose, organised a massive 600-bomber air raid in August 1943. Many

Left:
The distinct checker-patterned silhouette of an A-4 (V-2) rocket launching from the base.

Right:
RAF reconnaissance images of Peenemünde's *Prüfstand VII*, which hosted the first successful A-4 (V-2) launch before and after the carpet-bombing campaign against Germany's rocket island.

Bottom:
A V-1 falling from the sky over London. During the war, around 2,400 V-1s penetrated the UK's air defence system to instil a constant fear on its urban residents. The V-1, 'V' for 'vengeance is often associated in time with the V-2, but was quite different. The A-4, given the wartime title of the V-2, was a modern rocket, whereas the Fieseler 103, given the wartime title of V-1, was not a rocket at all, but a pulse jet, which uses a simple, low thrust repeat combustion jet engine'.
Source: USAF Archive

labourers and several leading rocket engineers were killed, notably Gerhard von Thiel, but key facilities were undamaged. Afterwards, testing was moved far inland to occupied Poland (Blizna) and production went underground (Nordhausen in the Harz mountains). Military A-4s were fired at Britain and continental targets from mobile stands from September 1944 to March 1945.

During the period of the German Democratic Republic (GDR; 1948–1990), Peenemünde became a military base and was deemed off-limits. Nowadays, it is open to visitors who can explore *Prüfstand VII* in the forest, the museum, and even climb into a Soviet-built U-boat that used to serve the GDR's navy.

On 17 August 1943, the Royal Air Force conducted a devastating air raid on the rocket facilities on the island.

Left:
A labelled map used by the RAF in its target selection.

Top:
The aftermath of the raid showing the wrecked *Prüfstand* VII.

Source: Royal Airforce Archive

After the tide of the air war had turned in the Allies' favour and following the bombing of Peenemünde, the focus shifted to mobile rocket batteries. In sync with the change of rocket doctrine, the bright checkered pattern gave way to more conventional camouflage.
Source: Bundes Archiv / USAF Archive

Launch of a V-2 rocket from Peenemünde.
Source: Bundes Archiv / USAF Archive

A mobile V-2 battery deployed in a forest. Despite the 70-year time difference, it does not fundamentally differ from today's military launch systems.
Source: Bundes Archiv / USAF Archive

During the early stages of the war, engineers could still make use of fully equipped test stands.
Source: Bundes Archiv / USAF Archive

Wernher von Braun surrenders to American forces in Oberammergau, Bavaria, on 2 May 1945.
Source: NASA

Left:
1954: Von Braun was visited by Walt Disney himself just nine years later. Together they produced three films about the future of space exploration with the aim of fuelling US national space ambitions.
Source: NASA

Some 21 days prior to von Braun's capture, US troops liberated the Nordhausen/Mittelbau-Dora concentration camp that had used slave workers to manufacture rockets in its underground factory up to that day.

In disbelief of the gruesomeness of the camp, US troops forced the local population to dig out mass graves for the dead prisoners left in the camp.
Source: Bundesarchiv

Today, the Thuringia tunnels and remains of rockets are part of European industrial history and are marketed as a tourist attraction.
Source: European Route of Industrial Heritage e.V.

A V-1 production line in the tunnels of Nordhausen, Thuringia (1944).
Source: Bundesarchiv

Germany in its current borders. The sites indicated show places in the context of the rocket programme that the Nazis would have enabled to develop the first nuclear weapon. Peenemünde was the main rocket assembling and testing site until 1943 when a British attack made further activities impossible.
Source: Paul Meuser

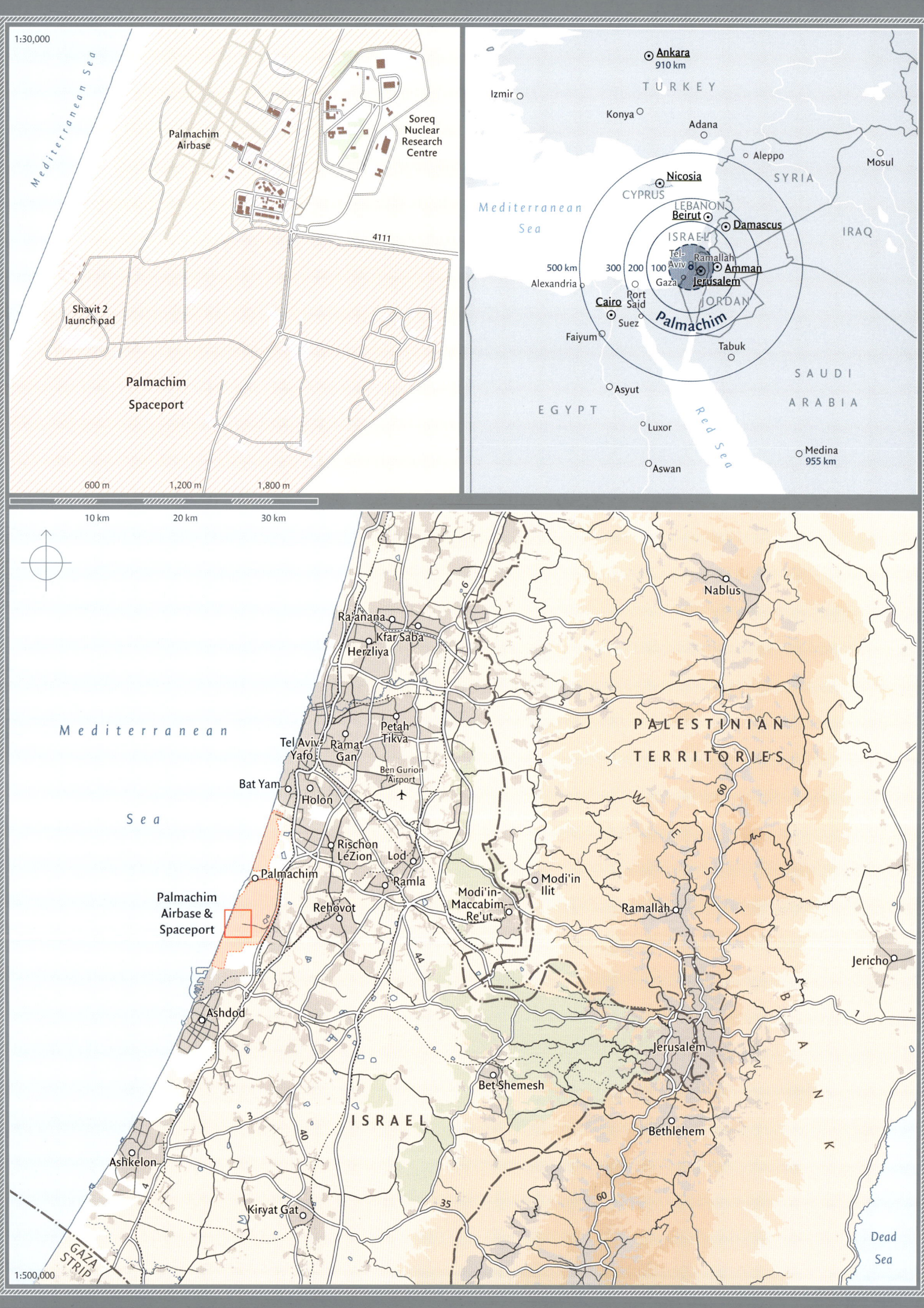

Inset map 1 (top left) — 1:30,000

Palmachim Airbase

Soreq Nuclear Research Centre

4111

Shavit 2 launch pad

Palmachim Spaceport

Mediterranean Sea

600 m 1,200 m 1,800 m

Inset map 2 (top right)

Ankara
910 km

Izmir

Konya

Adana

Aleppo

Mosul

TURKEY

SYRIA

IRAQ

Nicosia

CYPRUS

Mediterranean Sea

Beirut

LEBANON

Damascus

ISRAEL

Tel-Aviv

Ramallah

Amman

Jerusalem

JORDAN

Gaza

Palmachim

500 km 300 200 100

Alexandria

Cairo

Port Said

Suez

Faiyum

EGYPT

Asyut

Luxor

Aswan

Red Sea

Tabuk

SAUDI ARABIA

Medina
955 km

Main map (bottom) — 1:500,000

10 km 20 km 30 km

Mediterranean Sea

Nablus

Ra'anana

Kfar Saba

Herzliya

Petah Tikva

Tel Aviv-Yafo

Ramat Gan

Ben Gurion Airport

Bat Yam

Holon

PALESTINIAN TERRITORIES

Rischon LéZion

Lod

Ramla

Modi'in-Maccabim-Re'ut

Modi'in Ilit

Ramallah

Palmachim

Rehovot

Jericho

Palmachim Airbase & Spaceport

Ashdod

Jerusalem

Bet Shemesh

ISRAEL

Bethlehem

WEST BANK

Ashkelon

Kiryat Gat

GAZA STRIP

Dead Sea

1:500,000

ISRAEL'S BEACHSIDE MEDITERRANEAN LAUNCH SITE

Name
Palmachim Airbase

Location
Palmachim, Israel

Owner/Operator
Israeli Air Force, Israel Space Agency

Elevation
6 m

Coordinates
31.5°N, 34.6°E

Time zone
GMT+3

Launches
9 satellites

Completion
1988

Israel astonished the world by launching a small, 156-kg military observation satellite demonstrator, Ofeq, into orbit on 19 September 1988. Even more surprising was the pathway of its Shavit rocket, heading out westward over the Mediterranean Sea toward the Straits of Gibraltar, avoiding angry neighbours who might mistake it for a missile. Israel's space programme is, in global terms, unusual. It is almost entirely geared to meeting military needs, watching hostile countries (e.g., Syria, Iran), and keeping an eye on others (e.g., Egypt). It is one of only two launch sites in the Middle East, the other being Semnan (qv), Iran. Israel's space programme dates to sounding rockets in 1961. In the late 1970s, air-based reconnaissance became too risky, so the decision was made to develop an indigenous, high-tech space programme,

seen in the formation of the Israel Space Agency in 1983, its leader being Yuval Ne'eman. Its task was to develop both a launcher, Shavit, and a small military observation satellite, Ofeq. Shavit was a small, three-stage solid propellant rocket, weighing 23.4 tonnes and measuring 15.4 m in height. It was derived from the Jericho missile, using a mobile transporter. The launch site, Palmachim, is a two-runway air force base now used for helicopters and drones, just inland from beaches and an hour's drive south of Tel Aviv. It is named after Palmachim kibbutz, established there in 1949, and is close to the ancient seaport and archaeological site of Yavne. This is a closed air force base, not open to visitors, except the occasional Hollywood star at the invitation of the Israel Defence Forces. Little is known of the facilities there.

A SpaceX Falcon 9 rocket at Cape Canaveral
Air Force Base in 2019. On board is a lunar
lander named 'Beresheet', which was
developed by Israel's SpaceIL.
Source: Picture Alliance / newscom /
JOE MARINO - BILL CANTRELL

Right:
Despite hitching rides on commercial
flights for scientific missions, Israel continues to play its cards close to its chest when
it comes to national security. The Ofeq 16
spy satellite launches from the Palmachim
air base in central Israel 6 July 2020.
Source: Israel Ministry of Defence

Photography is discouraged, although people in the area can catch glimpses over the dunes and cannot help but notice contrails when rocket launches take place by day and their flames when they launch at night. Information on the Ofeq satellites is secret, although they are tracked by amateur observers. Israel used Palmachim for subsequent launches of Shavit and its successor Shavit 2 of Ofeq 2, 3, 5, 7, 9 10, and 11 (three failed). These have ever-improved imaging capacities, higher resolution, side-look photography, radar, and rapid data transmission. Launching westward has a high payload penalty, so larger Israeli satellites have also been launched by other countries (e.g., Russia, European countries, and India).

Declassified images of Syria's
Presidential Palace and international airport in Damascus taken
by Israel's Ofeq 11 spy satellite
launched on 13 September 2016.
The images were released two
years later on 17 September 2018.
Source: Israel Defence Ministry

Right:
Cover of *Maariv* on 20 September 1988. On the previous day, Israel entered the space age with the launch of its Ofeq 1 satellite. The satellite was active for 118 days and performed a number of solar cell and radio transmission tests.
Source: The National Library of Israel

Bottom:
'Don't shoot your ballot into space': During the 1961 elections, the Mapai party used the Shavit 2 sounding rocket launch on its political posters.
Source: The National Library of Israel

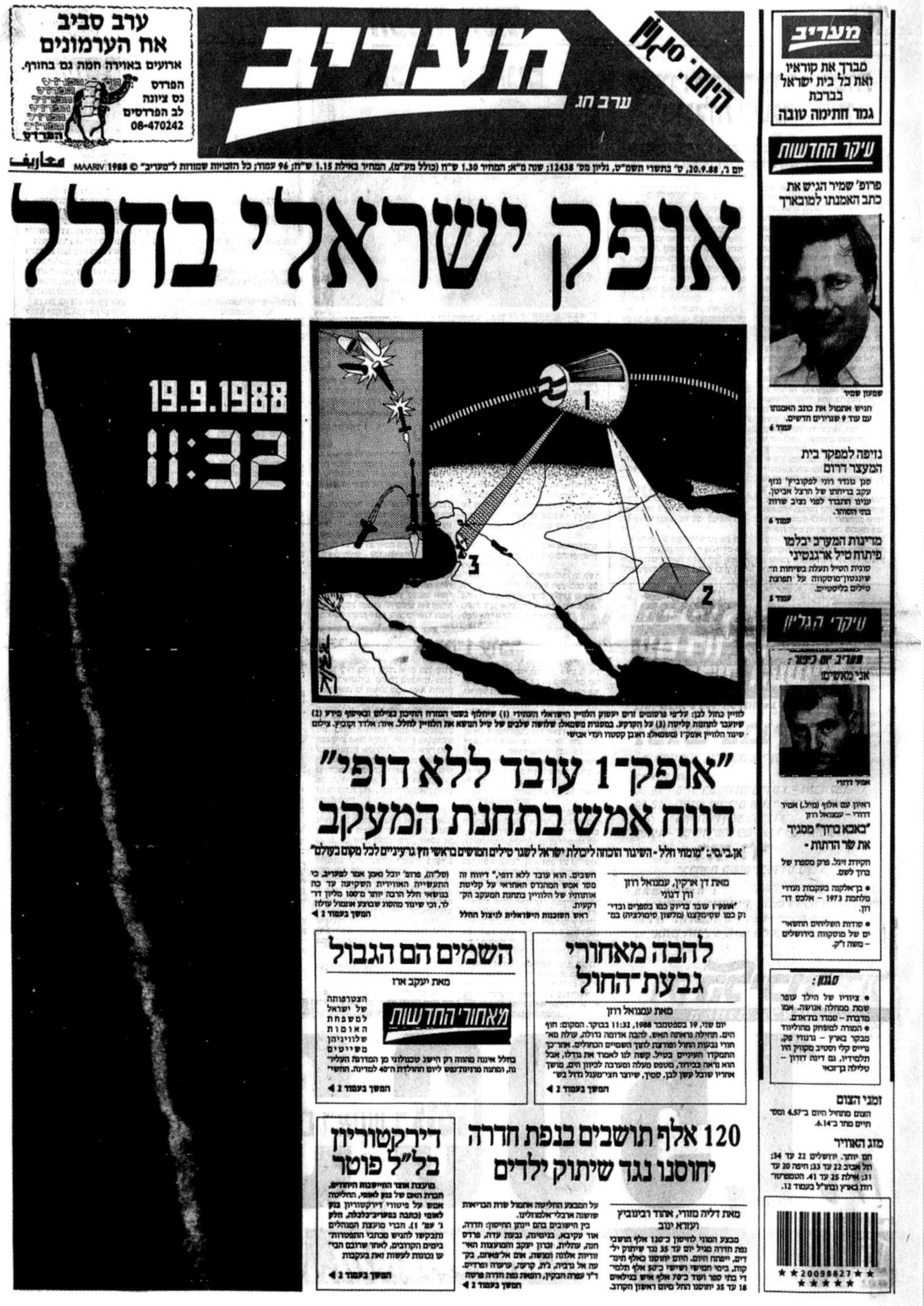

ערב סביב
אח הערמונים
ארועים באוירה חמה גם בחורף.
הפרדס
נס ציונה
לב הפרדסים
08-470242

מעריב
ערב חג
היום: סגנון

MAARIV 1988 ©

אופק ישראלי בחלל

19.9.1988
11:32

"אופק־1 עובד ללא דופי"
דווח אמש בתחנת המעקב

מאת דן ארקין, עמנואל רוזן
ודן דגוני

להבה מאחורי גבעת־החול

מאת עמנואל רוזן

המשך בעמוד 2

השמים הם הגבול

מאת יעקב ארז

מאחורי החדשות

המשך בעמוד 2

120 אלף תושבים בנפת חדרה יחוסנו נגד שיתוק ילדים

מאת דליה מזורי, אהוד רבינוביץ
ועזרא ינוב

המשך בעמוד 2

דירקטוריון בל"ל פוטר

המשך בעמוד 2

מעריב
מברך את קוראיו
ואת כל בית ישראל
בברכת
גמר חתימה טובה

עיקר החדשות

עיקרי הגליון

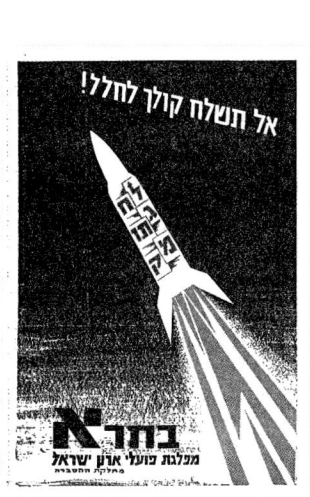

EURASIA

KAZAKHSTAN

RUSSIA

Baikonur ⊙

Plesetsk ⊙

Kapustin Yar ⊙

Caspian Sea

Sem ⊙

⊙

Baltic Sea

Black Sea

Palmachim ⊙

Peenemünde ⊙

Mediterranean Sea

Hammaguir ⊙

3

Sriharikota

Kulasekharapatnam

Thumba

IRAN

Persian
Gulf

Indian

Ocean

Equator

Red Sea

San Marco

Atlantic

Ocean

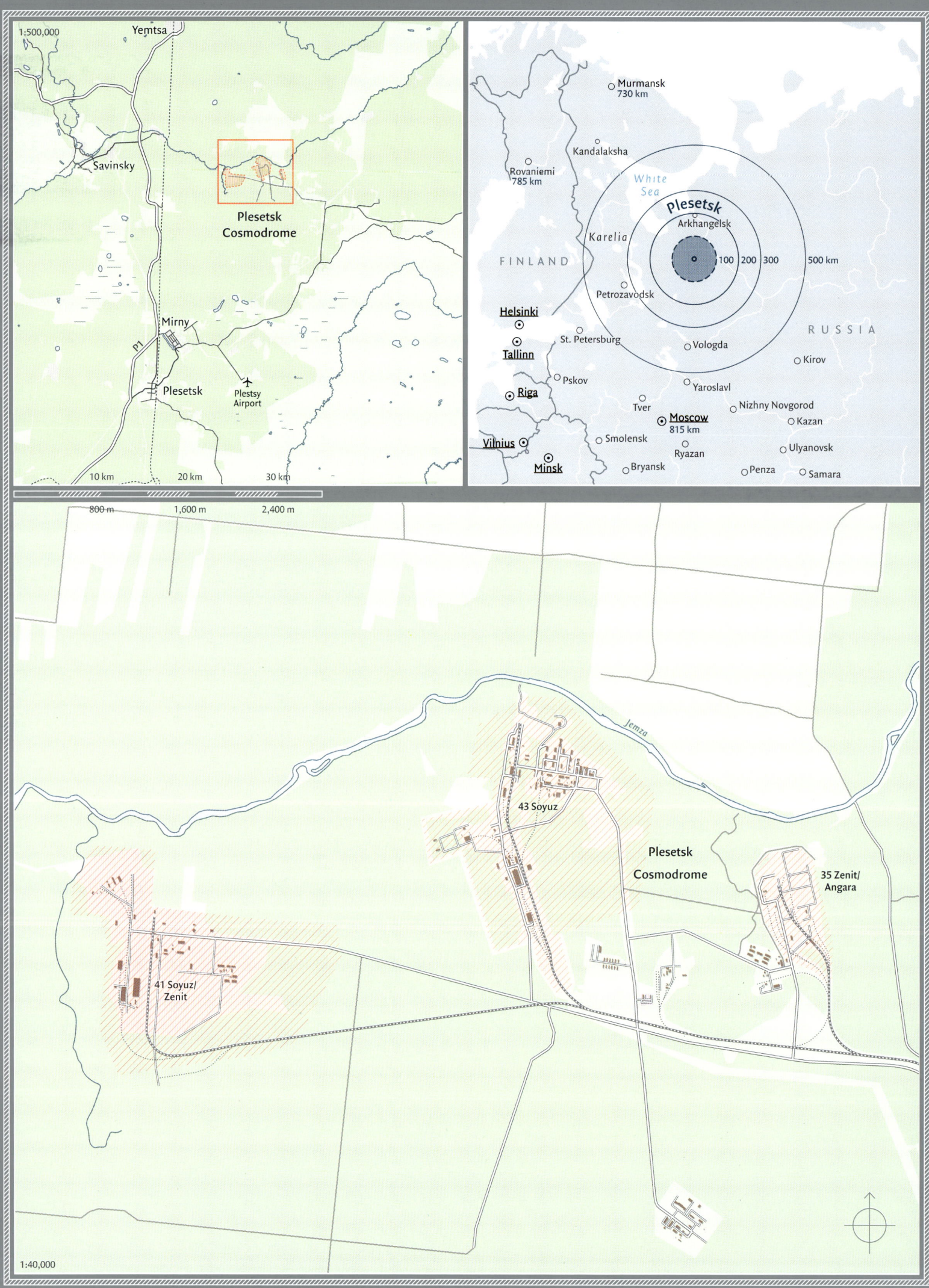

1:500,000

Yemtsa

Savinsky

Plesetsk
Cosmodrome

Mirny

P1

Plesetsk

Plestsy
Airport

10 km   20 km   30 km

Murmansk
730 km

Kandalaksha

Rovaniemi
785 km

*White
Sea*

Plesetsk

Arkhangelsk

Karelia

FINLAND

100 200 300   500 km

Petrozavodsk

Helsinki
⊙

Tallinn
⊙

St. Petersburg

Vologda

RUSSIA

Kirov

Pskov

Yaroslavl

Riga

Tver

Nizhny Novgorod

Kazan

Vilnius ⊙

Moscow
815 km

Smolensk

Ryazan

Ulyanovsk

Minsk

Bryansk

Penza

Samara

800 m   1,600 m   2,400 m

Jemza

43 Soyuz

Plesetsk
Cosmodrome

35 Zenit/
Angara

41 Soyuz/
Zenit

1:40,000

BUSIEST COSMODROME IN THE WORLD

Name		Coordinates
Plesetsk Cosmodrome		40.5°N, 62.8°N
Location		Time zone
Plesetsk, Russia		GMT+6
Owner / Operator		Launches
Russian Ministry of Defence		2,156 satellites
Elevation		Completion
85 m		1966

Given that more rockets have been launched from here than anywhere else, Plesetsk is surprisingly little known. Although Baikonur (qv) was and is Russia's primary cosmodrome, Plesetsk was the busiest and by 2000 had accounted for 38 per cent of all the world's launches.

The reason lies in its military origins and because it has since been used principally for military launches, though important scientific missions have also taken place from here. Plesetsk was not visited by non-Russians until the arrival of visitors from the socialist countries at the end of the Soviet period, with western Europeans following afterwards. Unlike Baikonur and Vostochny (qv) cosmodromes, which are civilian, Plesetsk remains under military control: launch workers are military staff and turn out in drill formation for inspection when rockets are being prepared. Plesetsk is in a forest area in the far north at 63°N, just short of the Arctic Circle (66°N) and 200 km south of Archangel. It is bitterly cold in winter and dark almost all day, while summers are warm with a short twilight and no real darkness. Temperatures are even more extreme than Baikonur, dropping as low as -46°C and routinely hovering around -20°C in midwinter, though this has never affected launches. It is the cosmodrome closest to western borders: space enthusiasts in Finland and Sweden have even seen the red and yellow glow of rockets rising into the distant night sky to the east.

Plesetsk started life as a military base for the Soviet Union's Intercontinental Ballistic Missile (ICBM) strike force. On 11 January 1957, General Secretary Nikita Khrushchev signed the order for the construction of the cosmodrome. The area comprised a mixture of dense forest, swamp, ravines, and rocky outcrops.

A heavy-class carrier rocket Angara-A5 is transported through the snow by train from its hangar to the launch pad.
Source: Russian Defence Ministry Press Service

Left:
The Soyuz-2.1b rocket booster belongs to an expendable rocket family dating back to 1966. Between the final flight of the space shuttle in 2011 and 2020 when SpaceX's Falcon 9 carried its first crewed mission to the ISS, the Soyuz rocket was the only launch vehicle capable of transporting astronauts to the space station.
Source: Picture Alliance / dpa / Sergei Medvedev

Lifting the almost 800,000 kg rocket from its means of transportation into its upright launch position is a technical feat in itself.
Source: Russian Defence Ministry Press Service

Next double page:
The luscious green forests in tandem with the disproportionate metal structures give Plesetsk an uncanny model scale appearance.
Source: Shutterstock (kaiser-v)

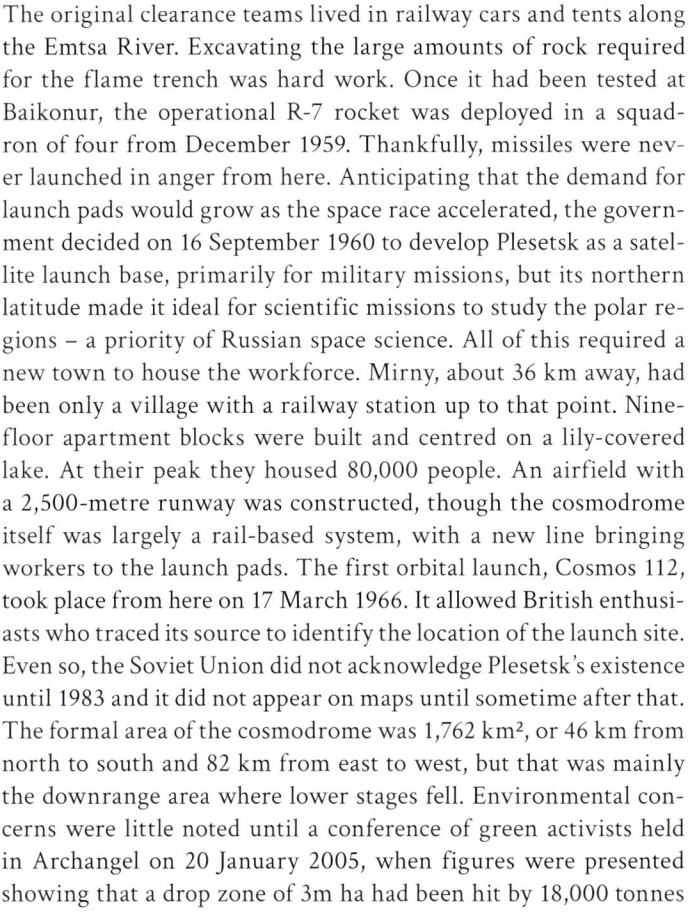

The original clearance teams lived in railway cars and tents along the Emtsa River. Excavating the large amounts of rock required for the flame trench was hard work. Once it had been tested at Baikonur, the operational R-7 rocket was deployed in a squadron of four from December 1959. Thankfully, missiles were never launched in anger from here. Anticipating that the demand for launch pads would grow as the space race accelerated, the government decided on 16 September 1960 to develop Plesetsk as a satellite launch base, primarily for military missions, but its northern latitude made it ideal for scientific missions to study the polar regions – a priority of Russian space science. All of this required a new town to house the workforce. Mirny, about 36 km away, had been only a village with a railway station up to that point. Nine-floor apartment blocks were built and centred on a lily-covered lake. At their peak they housed 80,000 people. An airfield with a 2,500-metre runway was constructed, though the cosmodrome itself was largely a rail-based system, with a new line bringing workers to the launch pads. The first orbital launch, Cosmos 112, took place from here on 17 March 1966. It allowed British enthusiasts who traced its source to identify the location of the launch site. Even so, the Soviet Union did not acknowledge Plesetsk's existence until 1983 and it did not appear on maps until sometime after that. The formal area of the cosmodrome was 1,762 km², or 46 km from north to south and 82 km from east to west, but that was mainly the downrange area where lower stages fell. Environmental concerns were little noted until a conference of green activists held in Archangel on 20 January 2005, when figures were presented showing that a drop zone of 3m ha had been hit by 18,000 tonnes

of scrap metal, 744 tonnes of oxidizer, 652 tonnes of kerosene, and 340 tonnes of heptyl, while the cosmodrome took 10m m³ of water from the ground every year.

Like Baikonur, Plesetsk had its share of disasters, with eight fatalities from a launch explosion on 26 June 1973 and 51 rocket workers killed in another explosion on 18 March 1980. Secret at the time, they were marked by a memorial beside the lake and a custom of never launching rockets on those days again.

The original cosmodrome comprised four R-7 pads, called 43 (two), 16, and 41 (dismantled 1989) and used by its subsequent iterations (e.g., Vostok, Molniya, Soyuz), followed by two Cosmos pads (131, 132; now decommissioned), and two Tsyklon pads (32), decommissioned in 2001 when the Tsyklon went out of service. A Zenit pad (35) was completed in 1986, but never used and in 1999 the decision was taken to convert it into an Angara pad. Near Cosmos pad 132 is a Rockot pad, 133, but it is more like a tower to which the small Rockot launcher is brought by a military trailer. Compared to Baikonur, the launching area is quite compact – a feature of the difficult terrain. Near the collection of pads there are seven assembly shops and integration halls, including a large MIK with a huge oxygen and fuel plant – reportedly the biggest in Europe. In the 1980s, Plesetsk was so busy that launches were almost weekly.

Conditions deteriorated in Plesetsk from the early 1990s. For the rocket troops and soldiers on military service there, conditions were harsh, with run-down buildings and food limited largely to bread, gruel, soup, and eggs, supplemented by potatoes grown in the grounds of the general hospital. Their children were educated

The gantry arms begin to close around a Molniya-M booster rocket carrying a Kosmos military satellite into orbit.
Source: Picture Alliance / dpa / Tass Babenko Alexander

in prefabs, which is no joke in the Arctic. Wages went into arrears. The road network was in such an atrocious condition by 1992 that it nearly broke the chassis of Boris Yeltsin's limousine when he visited (he issued a presidential decree ordering improvements). They no longer even had a training rocket to use to instruct the rocket troops, who had less and less to do. Fewer and fewer military satellites were launched – an average of one every two months, paralleling the general decline of the programme. As was the case in Baikonur, salvation lay with western companies eager to use Russia's greatest space asset: its fleet of reliable rockets, which were much less expensive than Western ones. Eurockot invested €34m in a payload integration facility, clean rooms, mission control, and a hotel ('The Eurockot') and customer facilities in Mirny. Payloads were flown to Archangel and then brought 200 km by rail to Plesetsk. From the mid-1990s, there was a steady stream of Europeans and Americans arriving in Plesetsk with their cargoes, often flown on Cold War surplus missiles. With their cash, facilities in Plesetsk began to improve again. In 2001, a simulator arrived to train the rocket troops. Russian government funding finally arrived for an Angara pad in the mid-2000s, but it took 10 years to complete, with its first orbital launch being the Angara 5 – Russia's most powerful rocket – on 23 December 2014. The Angara is brought horizontally from the MIK by diesel train, erected to the vertical, and clasped by a big, green-painted 1,185-tonne launch tower, overseen by an underground launch centre 15 m downrange. Flying non-military satellites from Plesetsk, a military base, presented a problem. Scientists in the Soviet Union were well used to packing up and sending

A local using a hoe made from a rocket part to scavenge for usable parts among the space junk.
Source: Raffaele Petralla

Next double page:
What goes up must come down: metal scraps from fuel containers and rocket boosters stack up in a nearby village.
Source: Raffaele Petralla

away their instruments and that was the last they saw of them until they were informed that they were now in orbit. They were not involved in the integration process, which was procedurally risky. This rule was first relaxed under the Interkosmos programme when representatives of the socialist countries were permitted to enter Plesetsk. One Czechoslovakian scientist, there for a satellite launch in 1989, recalled how it was very controlled, 'with walls and military everywhere'. Western scientists repeatedly asked to be allowed into Plesetsk, an urgent imperative when biological cargoes required careful handling at their final stages. They were finally admitted for Bion 10 in 1992. Western companies like the German Eurockot and the British Commercial Space Technologies arrived in the mid-1990s: they were guaranteed that their customers' satellites would be under their supervision at all times, so the security rules had to be amended. Western journalists were allowed in for the Foton 10 launch in 1995. Eventually, the launches of European Space Agency satellites (e.g., Sentinel) from Plesetsk were televised live. The period of post-Soviet commercialisation may now have run its course, as a range of new, small Western launchers (e.g., Vega) are well able to handle the full range of European payloads, so Plesetsk has reverted to its earlier role as primarily a military launch base. In long-term Russian planning, Plesetsk is the military base, while Vostochny will be the civilian one. Current launches are typically military satellites such as Meridian (communications), Liana (electronic intelligence), Tundra (early warning), Persona (visual observations), Bars (mapping), as well as navigation satellites (GLONASS). Most are launched on the Soyuz 2 rocket, a descendant of the original R-7 flown from its original pads, so history has come full circle.

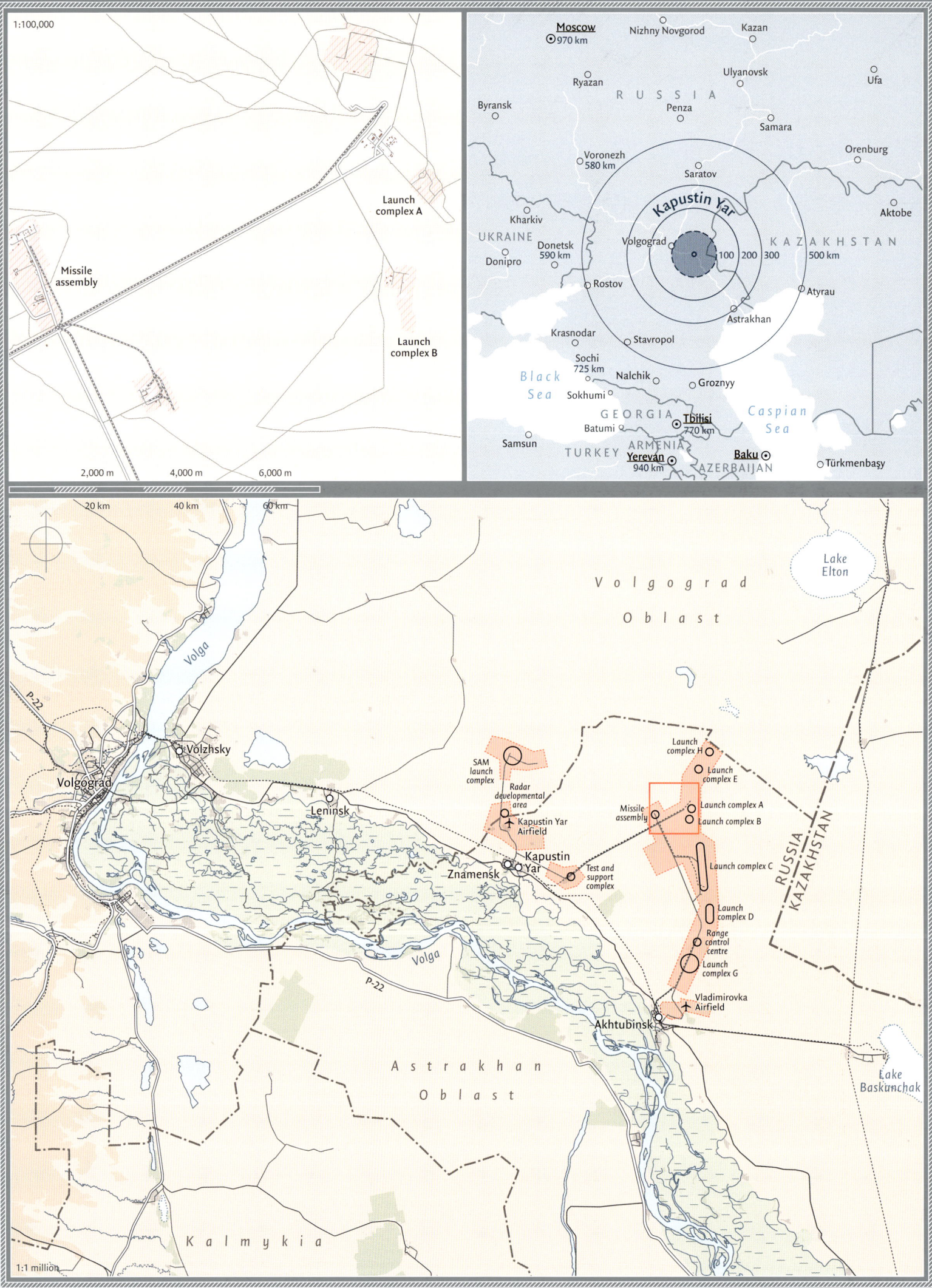

1:100,000

Launch
complex A

Missile
assembly

Launch
complex B

2,000 m   4,000 m   6,000 m

Moscow
⊙ 970 km   Nizhny Novgorod   Kazan

Ryazan   Ulyanovsk   Ufa
Byransk   Penza
                                          Samara
R U S S I A
Voronezh   Orenburg
580 km   Saratov
Kharkiv   Aktobe

UKRAINE   **Kapustin Yar**   K A Z A K H S T A N
Donipro   Volgograd
                            590 km   100 200 300   500 km
          Rostov   Atyrau

                        Astrakhan
Krasnodar   Stavropol
Sochi   *Caspian*
725 km   Nalchik   Groznyy   *Sea*
    Sokhumi
          G E O R G I A   *Black*
Samsun   Batumi   **Tbilisi**   *Sea*
TURKEY   770 km
          ARMENIA   **Baku** ⊙
              Yerevan ⊙   AZERBAIJAN   Türkmenbaşy
                940 km

20 km   40 km   60 km

*Lake
Elton*

**V o l g o g r a d
O b l a s t**

Volga

P-22

● Volzhsky

Launch
complex H
                    Launch
                    complex E
SAM
launch
complex   Missile   Launch complex A
Radar   assembly   Launch complex B
developmental
area   Kapustin Yar
          Airfield   Launch complex C
Kapustin
Leninsk   Yar
          Znamensk   Test and
                     support   Launch
                     complex   complex D
Volgograd
                              Range
                              control
                              centre
                              Launch
              Volga   complex G

          P-22
                              Vladimirovka
                              Airfield
                    Akhtubinsk

R U S S I A
KAZAKHSTAN

**A s t r a k h a n
O b l a s t**
                                        *Lake
                                        Baskunchak*

K a l m y k i a

1:1 million

RUSSIA'S ORIGINAL COSMODROME

Name		**Coordinates**
Kapustin Yar		48.4°N, 45.8°E
Location		**Time zone**
Kapustin Yar, Russia		GMT+4
Owner/Operator		**Launches**
Russian Aerospace Forces		92 satellites
Elevation		**Completion**
13 m		1946

Kasputin Yar was the Soviet Union's original cosmodrome, although the first orbital launch, Sputnik, was made from Baikonur. When Russian engineers found the German A-4, they needed a location to test it, so they settled on flat land to the east of Stalingrad, the famous battleground that was later renamed Volgograd. Railways brought the German A-4s there on flat wagons, while the engineers lived in converted railway carriages. The A-4 required little infrastructure: a concrete pad, an erector, fuelling systems, and tracking devices. The first A-4 was fired from here on 18 October 1947, followed by the R-1 (Raket, or rocket), the reverse-engineered Russian version, in 1948 (R-2), and larger suborbital rockets like the R-5. On 22 July 1951, the first dogs were launched 100 km into space from Kapustin Yar, which was used for dog and *Akademik* ('scientific') missions right through the 1950s. During these years, Royal Air Force Canberra bombers braved flak over Ukraine to overfly and photograph the launch site. In 1962, the USSR began the Cosmos programme of small scientific satellites, most being made by the Mikhail Yangel OKB-586 design bureau in Dnepropetrovsk in Ukraine to the west, as was his R-12 and subsequent R-14 missile, in civilian form called the Cosmos (2 ,2M 3, 3M versions). Kapustin Yar was geographically convenient and not difficult to adapt, so accordingly, Cosmos 1 was launched from a silo here on 16 March 1962, followed by other scientific satellites in the Cosmos programme. India's first satellite, Aryabhata, was launched from here in 1975. It was also Kapustin Yar that was used for tests of the small unmanned BOR (*Bespilotnyi Orbitalni Rakekoplan* or 'unpiloted orbital space plane') space plane over 1982 – 1984, which circled the Earth and splashed down in the Black Sea or Indian Ocean.

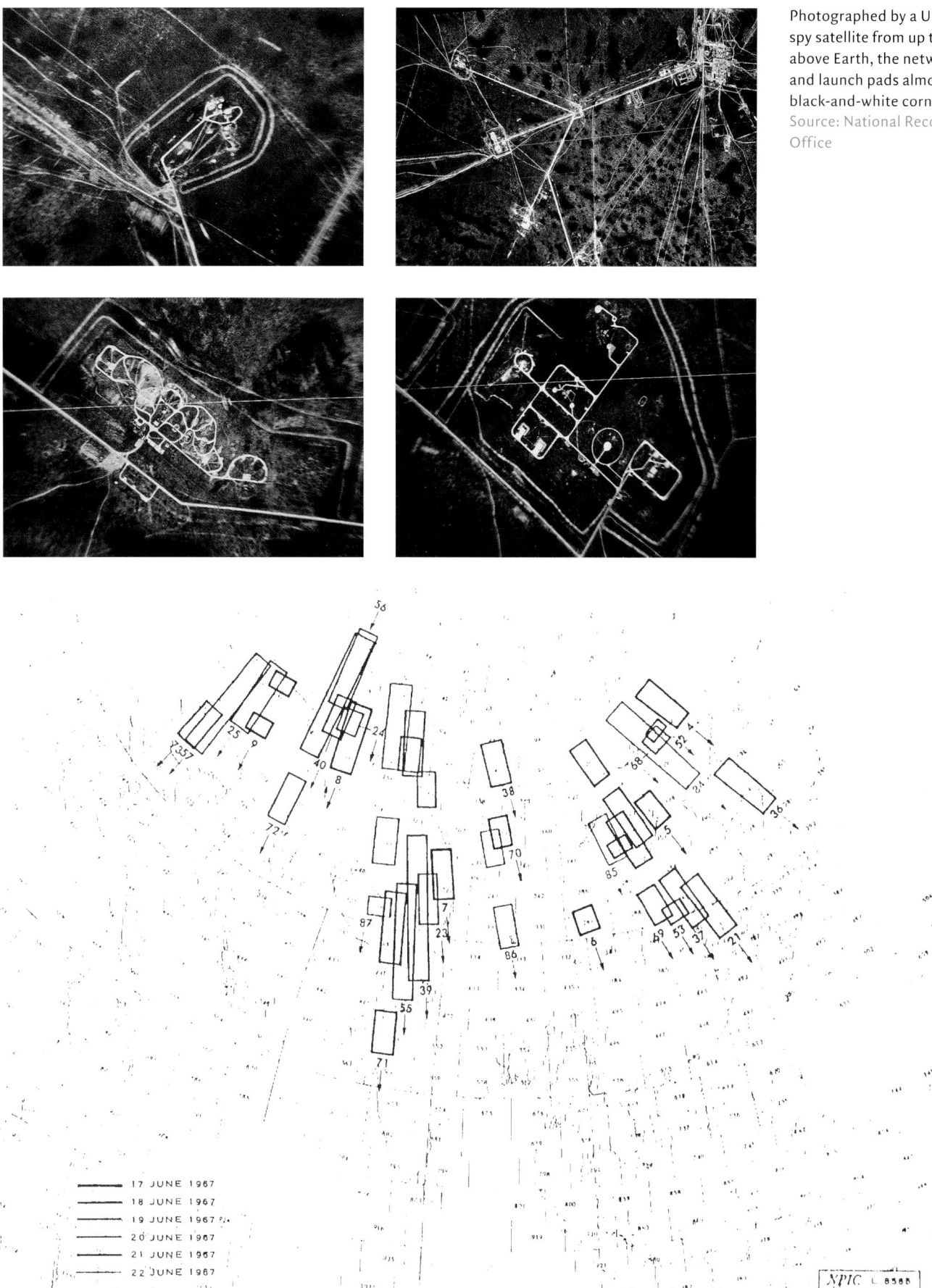

Photographed by a US Corona spy satellite from up to 460 km above Earth, the network of roads and launch pads almost looks like black-and-white corn circles.
Source: National Reconnaissance Office

17 JUNE 1967
18 JUNE 1967
19 JUNE 1967
20 JUNE 1967
21 JUNE 1967
22 JUNE 1967

Approximate single satellite coverage over the Soviet Union.
Source: National Reconnaissance Office

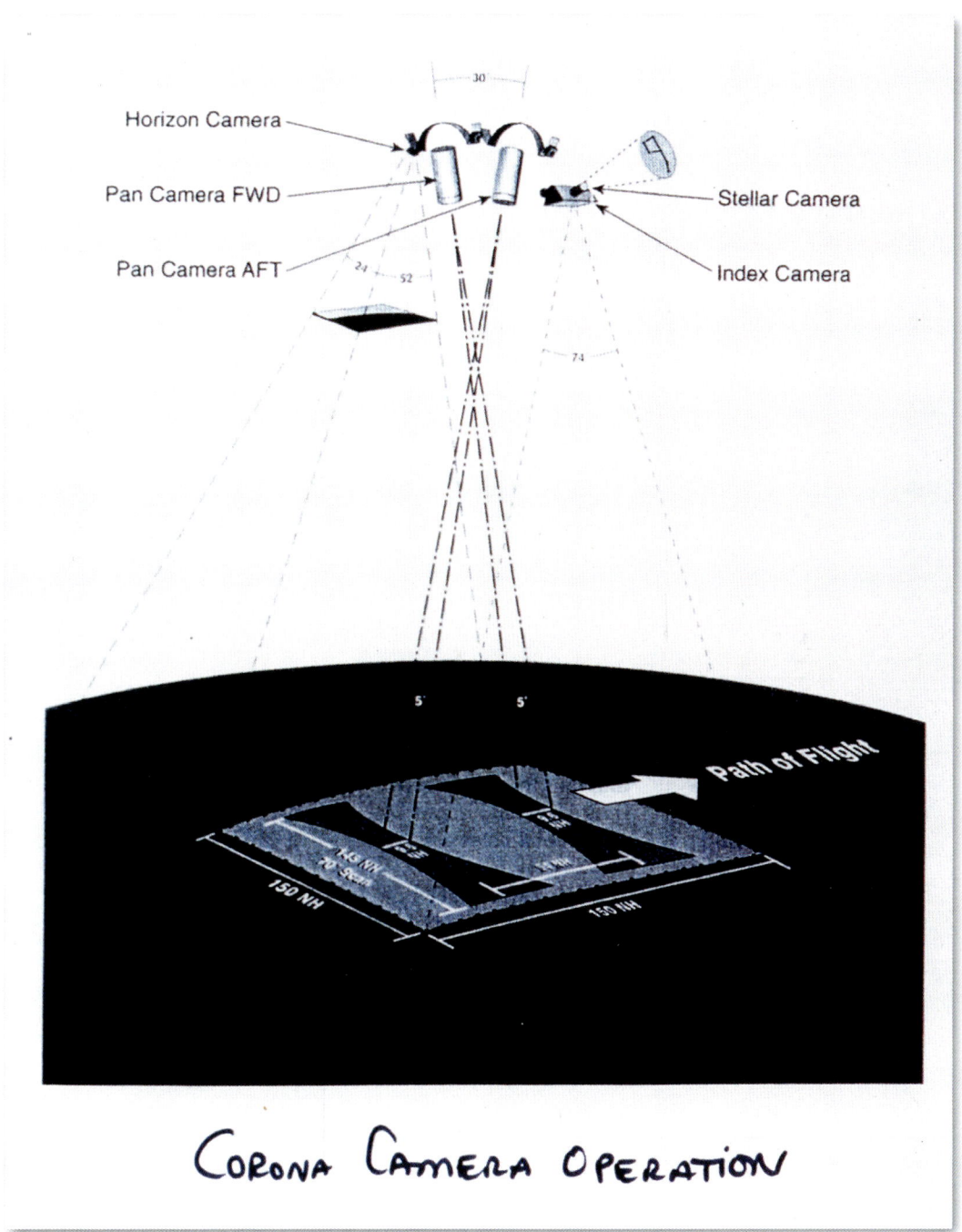

Horizon Camera

Pan Camera FWD

Pan Camera AFT

Stellar Camera

Index Camera

Path of Flight

CORONA CAMERA OPERATION

At the end of its mission, the satellite would eject a re-entry capsule, which would deploy its parachute and be caught in mid-air by a specially-adapted airplane.
Source: National Reconnaissance Office

The base then fell into disuse, but was used as a location for de-commissioning Cold War missiles in the 1990s. In January 1997, sounding rockets were fired from here as part of an American-Russian cooperative programme and President Yeltsin paid a visit to mark its 50th anniversary. April 1999 saw its last orbital launch, the German X-ray observatory Abrixas on a Cosmos 3M. The last launch was a Cosmos 3M suborbital military test launch on 22 April 2006 and the rocket then went out of production.

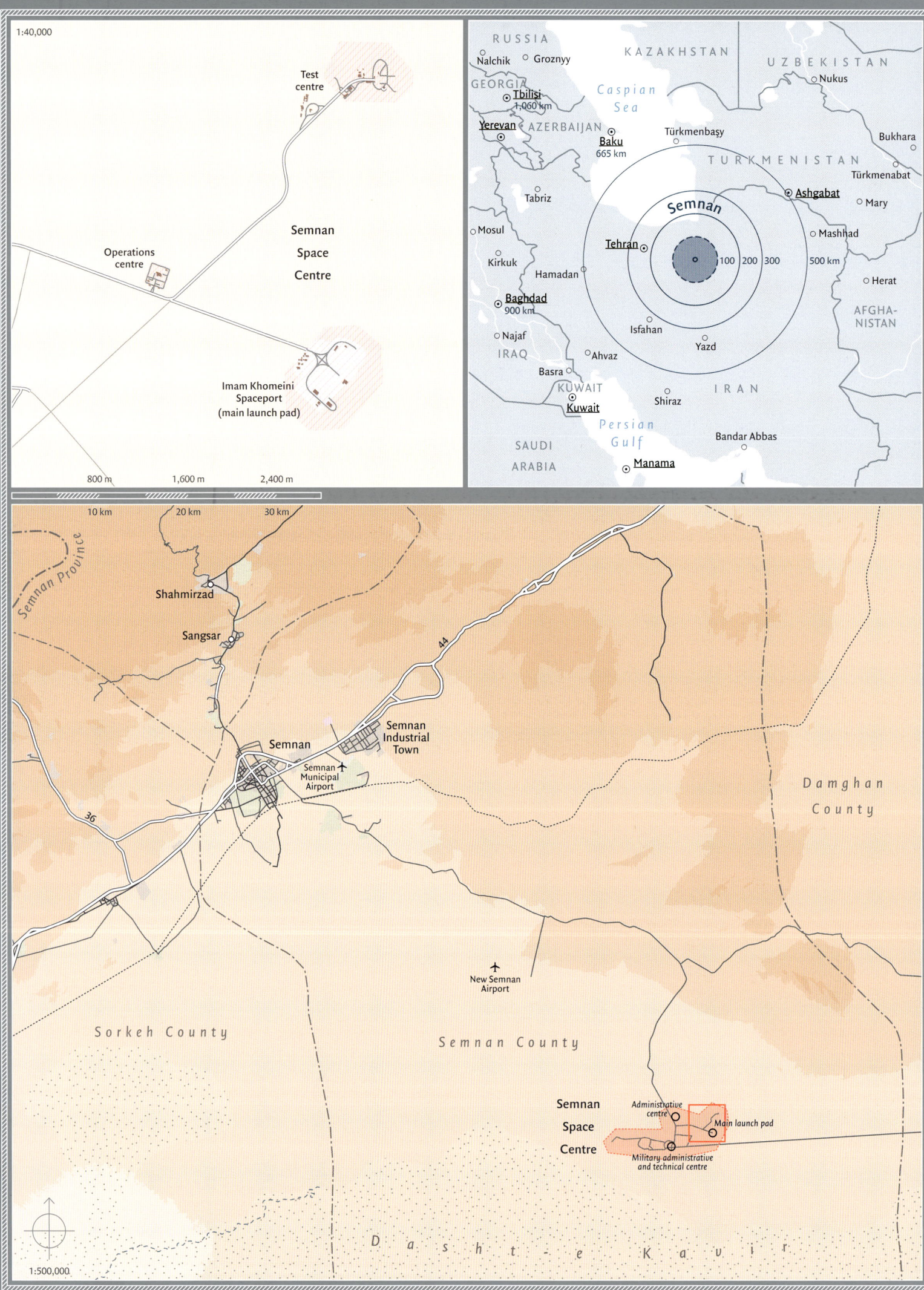

1:40,000

Test
centre

Semnan
Space
Centre

Operations
centre

Imam Khomeini
Spaceport
(main launch pad)

800 m   1,600 m   2,400 m

RUSSIA
Nalchik   Groznyy   KAZAKHSTAN   UZBEKISTAN
GEORGIA   Nukus
Tbilisi   Caspian
1,060 km   Sea   Türkmenbaşy   Bukhara
Yerevan   AZERBAIJAN   Baku   TURKMENISTAN
665 km   Türkmenabat
Tabriz   Ashgabat   Mary
Mosul   Semnan   Mashhad
Tehran
Kirkuk   Hamadan   100 200 300   500 km
Baghdad   Isfahan   Herat
900 km   Yazd   AFGHA-
Najaf   NISTAN
IRAQ   Ahvaz
Basra   IRAN
KUWAIT   Shiraz
Kuwait   Bandar Abbas
SAUDI   Persian
ARABIA   Gulf   Manama

10 km   20 km   30 km

Semnan Province

Shahmirzad

Sangsar

44

Semnan
Industrial
Town

Semnan

Semnan Municipal
Airport

Damghan
County

36

New Semnan
Airport

Sorkeh County

Semnan County

Semnan
Space
Centre

Administrative
centre

Main launch pad

Military administrative
and technical centre

D a s h t - e   K a v i r

1:500,000

IRAN'S IMAM KHOMEINI DESERT LAUNCH SITE

Name
Imam Khomeini
National Space Base

Location
Semnan, Iran

Owner/Operator
Iranian Space Agency

Elevation
1,100 m

Coordinates
35.2°N, 53.9°E

Time zone
UTC+06:00

Launches
5 satellites

Completion
2009

Iran is one of the least well-known space powers, orbiting its first satellite, Omid, on 2 February 2009. It was an early participant in international space activities, notably communications and remote sensing, developing an indigenous space programme from 2004. Iran's launch site is Semnan, which is also the name of the province where it is located. It is 170 km east of Tehran on the Dasht-e Kavir desert plateau, ringed by distant mountains. The site was originally used for sounding rockets (Kavoshgar), missiles (Shahab), and satellite (Safir) launchers. Its basic, circular concrete pad measured 60 m in diameter. The Safir rocket was brought there on a military-style transporter-erector with a fuelling gantry of six levels, which was driven away shortly before launch. There were no permanent facilities, only trucks to bring fuel, gas, and electrical power, with others serving as a command centre and for tracking. Semnan's existence was first made known through the much-publicised 200-km-high suborbital launch of the Kavoshgar 1 sounding rocket on 4 February 2008, attended by President Mahmoud Ahmadinejad. No foreign visitors have been there: our knowledge is limited to scenes shown on Iranian television and overhead imaging by Earth observation satellites. The Shahab, Kavoshgar, and Safir rockets are similar: small, long, and thin and derived from the Democratic People's Republic of Korea adaptations of the Soviet Scud missile from the 1950s. The two-stage liquid-fuelled Safir weighed 26 tonnes, was 22 m long, and 1.25 m in diameter, putting it at the minimalist end of global launchers. Omid itself weighed 27 kg and took a southbound path toward the Arabian Sea and Indian Ocean. Safir was used for suborbital launches of monkeys and for subsequent satellite

This satellite image from Maxar Technologies
shows a fire at a rocket launch pad at the Imam
Khomeini Space Center in Iran's Semnan province,
August, 2019.

An Iranian rocket carrying a satellite is launched
from an undisclosed site believed to be in Iran's
Semnan province. Iran's Revolutionary Guard
declared that the Islamic Republic's first military
satellite launch into orbit had taken place on
20 April 2020, dramatically unveiling what experts
described as a secret space programme with a
surprise launch that came amid wider tensions
with the United States.

launches: Rasad (2011), Navid (2012), and Fajar (2015). Another
site further to the north-east, Shahroud, may have been used
for the military Nour satellite (2020). There was a big expansion
of Semnan associated with the more powerful Simorgh rocket
(2016), which is able to orbit 250 kg. A new pad was built, with
a large white enclosed launch tower, gantry, four permanent service buildings, underground launch control facility, and four high
electrical masts. Roads were built and buildings for operations,
administration, and rocket engine testing were installed. It was
named the Imam Khomeini Launch Centre after the leader of the
Islamic Revolution (1979). His image is imprinted on the launch
tower. Several launch attempts on Simorgh rockets have failed or
been sabotaged.

1958: After the launch of humanity's first artificial satellite, Sputnik, the still young UN sprang into action forming an ad hoc Committee on the Peaceful Uses of Outer Space (COPUOS), composed of 18 members including Iran.
Source: United Nations Office for Outer Space Affairs

2016: An image released by the Iranian Defence Ministry showing the launch of a satellite-carrying rocket from the gantry at Imam Khomeini National Space Base.
Source: Iranian Defence Ministry

2021: The launch of Iran's newest
satellite-carrier rocket, 'Zuljanah'.
Source: Iranian Defence Ministry

2009: Iran launched its domestically-
produced satellite rocket, Safir Omid
('Hope Envoy').
Source: Iranian Defence Ministry

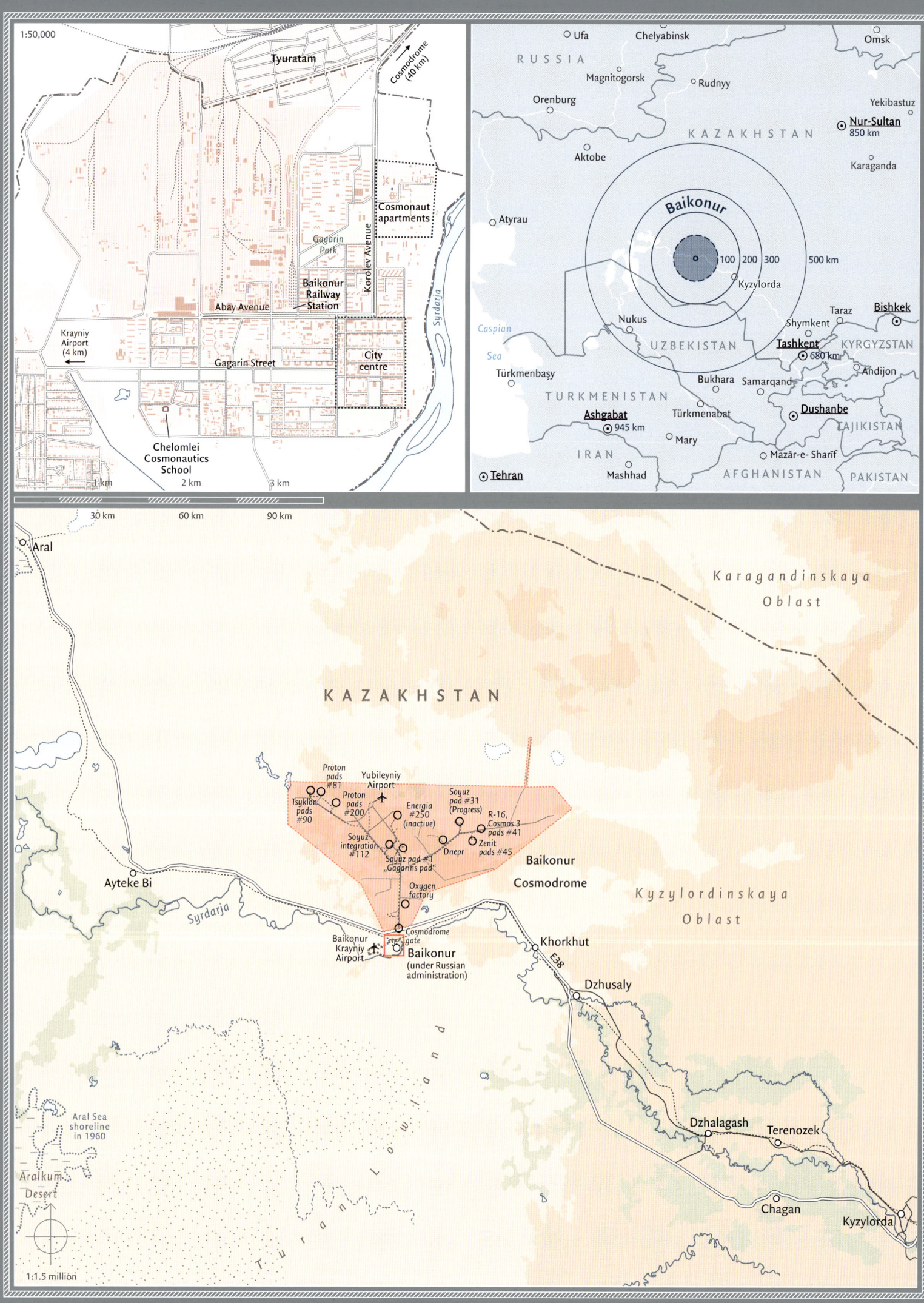

Inset map 1 (top left) — City map, 1:50,000

Tyuratam

Cosmodrome (40 km) →

Cosmonaut apartments

Gagarin Park

Korolev Avenue

Baikonur Railway Station

Abay Avenue

Syrdarja

Krayniy Airport (4 km) ←

Gagarin Street

City centre

Chelomlei Cosmonautics School

1 km 2 km 3 km

Inset map 2 (top right) — Regional overview

RUSSIA

Ufa Chelyabinsk Omsk

Magnitogorsk Rudnyy Yekibastuz

Orenburg

KAZAKHSTAN

Nur-Sultan 850 km

Aktobe Karaganda

Atyrau

Baikonur

100 200 300 500 km

Kyzylorda

Taraz Bishkek

Nukus Shymkent

Caspian Sea UZBEKISTAN Tashkent 680 km KYRGYZSTAN

Türkmenbaşy Bukhara Samarqand Andijon

TURKMENISTAN

Ashgabat 945 km Türkmenabat Mary Dushanbe TAJIKISTAN

IRAN Mashhad Mazār-e- Sharīf AFGHANISTAN PAKISTAN

Tehran

Main map — Baikonur Cosmodrome region, 1:1.5 million

30 km 60 km 90 km

Aral

KARAGANDINSKAYA Oblast

KAZAKHSTAN

Proton pads #81

Yubileyniy Airport

Proton pads #200

Tsyklon pads #90

Energia #250 (inactive)

Soyuz pad #31 (Progress)

R-16, Cosmos 3 pads #41

Soyuz integration #112

Zenit pads #45

Dnepr

Soyuz pad #1 "Gagarins pad"

Baikonur Cosmodrome

Kyzylordinskaya Oblast

Ayteke Bi

Syrdarja

Oxygen factory

Cosmodrome gate

Baikonur Krayniy Airport

Khorkhut

Baikonur (under Russian administration)

E38

Dzhusaly

Aral Sea shoreline in 1960

Turan Lowland

Dzhalagash Terenozek

Aralkum Desert

Chagan Kyzylorda

WHENCE SPUTNIK AND GAGARIN LEFT THE EARTH BEHIND

Name
Baikonur Cosmodrome

Location
Baikonur, Russia

Owner / Operator
Roscosmos State Corporation
Russian Aerospace Forces

Elevation
100 m

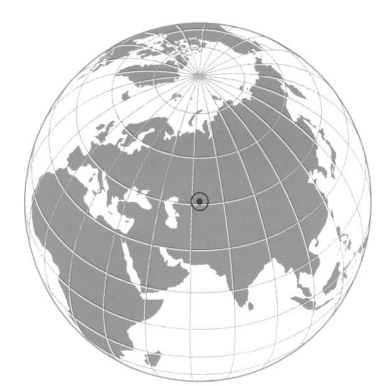

Coordinates
45.6°N, 63.4°E

Time zone
GMT +6

Launches
1,979 satellites

Completion
1957

Baikonur will forever be known as the cosmodrome from where the world's first satellite, Sputnik, was launched in 1957 and Yuri Gagarin made the world's first spaceflight four years later in 1961. Baikonur has been through expansion, contraction, decline, rebuilding, and political turmoil; even its future is uncertain. Weatherwise, Baikonur is a place of extremes, from freezing cold in winter – as low as -30°C – to roasting temperatures up to 40°C in the summer. It is at its most pleasant and colourful in spring when small flowers blossom all over the desert. The origins of Baikonur lie in the Soviet Union's need for an intercontinental ballistic missile (ICBM) to deliver an atomic warhead to the other side of the world, though its designers, Sergei Korolev and Mikhail Tikhonravov, were well aware that their rocket, called the R-7, could also launch satellites. Testing such a military missile required secrecy and because Kapustin Yar (qv) was already known to the West, a desert site far inland in Kazakhstan was chosen. The cosmodrome was called Baikonur after a village near a quarry on a railway line 350 km away, in the hope that Western missiles would be mistakenly targeted there rather than the real location, which was on a bend in the Syr Darya River alongside the Moscow-Tashkent railway, locally called Tyuratam. American U-2 spy planes found the real cosmodrome quickly enough. The facilities stood out against the desert landscape, especially in the snow, while the ICBM required a launch pad on pillars on a readily visible enormous excavated flame trench. The large assembly building, called the MIK, located close by and completed 1956, also stood out. Rockets were brought from here flat on the back of diesel-pulled railway cars. Construction began in summer 1955.

Scaffolding allowing engineers to work on the horizontally-assembled rocket stages.
Source: DOM publishers Archive

Rail tracks leading into one of the large assembly buildings.
Source: DOM publishers Archive

Right:
Buran, the Soviet space shuttle on its launch pad.
Source: DOM publishers Archive

From early on, it became the largest cosmodrome in the world. With the site measuring 90 km by 75 km, it covered an area of 6,717 km², extending far into the desert north-east from its residential area, originally called Leninsk and subsequently renamed Baikonur. Transport here is largely railway-based, with 470 km of track. Rockets, stages, and satellites arrive by train from Moscow and Samara, with an airfield, Krainy, used for personnel and some cargo. There is a 1,281-km road network, but it is bumpy and not always well maintained. The downrange fall zone is 104,305 km² – as large as some small countries. The launch pad completed in 1957 was the classic shape of four petals that clamped the rocket, called the tulipan ('tulip'), set over a complex bear-trap installation that held the base of the rocket and the huge flame trench. As the engines built up thrust, the lift exceeds the weight of the clamps, which fall back mechanically under gravity and release the rocket. This was pad 1, subsequently renamed Gagarinsky Start after the first piloted space mission and was the main point of departure for subsequent variants like the Soyuz. It was in almost continuous use until 2019 when it was retired. Launch control was in an underground bunker accessible nearby through a metallic door, with a periscope used to observe the launch itself. Launch was commanded by turning a key. Despite its apparent simplicity, it was a very capable system, with the same pad able to launch cosmonauts on Vostok 3 and Vostok 4 only 24 hours apart in 1962.

A second such pad (pad 31) was built 30 km to the east in 1961 as an R-7 missile pad, but then used for standard space missions from 1964, with its own assembly building called MIK 40. In the new century it was converted to the latest iteration of the R-7, the computer-controlled Soyuz 2 rocket (no more launch keys). There are no visible or verbal countdown clocks, a legacy of the days when American eavesdroppers used to listen in to upcoming launches from their bases in Turkey. There was a substantial expansion of Baikonur for the Moon programme in the 1960s. Firstly, about 35 km north-west of pad 1, a set of four pads for the powerful UR-500K Proton rocket was constructed to send cosmonauts around the Moon. These pads were built in two neighbouring complexes, 81 and 200, only 600 m apart and flanked by two 110-metre-high towers combining the functions of lightning conductor, TV camera point, and floodlamp location. Proton had a 50-metre launch tower though which cosmonauts would board their Moonship and which would be rolled back before launch. Proton's orange and brown flame was carried away by horizontal trenches. Located several kilometres away was a low horizontal assembly and integration facility measuring 120 m in length and 50 m in width. It was able to hold up to six full Protons at a time, each delivered by train from the Khrunichev factory in Moscow. Proton first flew in 1965, but when the crewed lunar programme was abandoned, Proton was used to launch robotic lunar and interplanetary spacecraft, communications satellites, and space stations. Secondly, construction began to the north-west, not far from the original R-7 pad, of two huge pads for the N-1 rocket intended to land cosmonauts on the Moon. It had its own 250-metre-long assembly hangar – about the size of old Zeppelin airship sheds – with giant railway-based transporters. Four N-1s were launched, none successfully. After the N-1 was cancelled in 1974, they lay idle until the 1980s when they were converted for the enormous Energiya rocket, which first flew in 1987 and the following year launched the space shuttle Buran before it too was cancelled. On its single, uncrewed flight, Buran returned to a

Left:
Energiya towering over the dry
steppe. The flat plains around
Baikonur offer an endless site for
the ever expanding
launch complex.
Source: DOM publishers Archive

Right:
As recognisable as the Soyuz
rockets themselves, their
launch towers are distinct in
their silhouette and appearance.
Source: DOM publishers Archive

nearby 84-metre wide, 4.5-km-long runway constructed with polished high-grade concrete, now called *Yubeleniye* ('anniversary airfield'). Energiya's first stage, the Zenit, was a powerful rocket in its own right, with its own complex of double sites to the south-east, pad 45 (one was destroyed in a launch accident). Missiles were also tested in Baikonur and in 1960, an R-36 rocket exploded, killing over 91 people, the world's worst launch disaster. It became known as the Nedelin disaster after the Marshal in charge, Mitrofan Nedelin. In Baikonur's heyday in the 1980s, Tsyklon and Cosmos rockets took off from other sites.

Baikonur suffered greatly from the contraction of the programme that began at the end of the Soviet period, a low point being the collapse of the building that housed Buran. When the new Russian space programme attracted commercial Western funding, new facilities and hotels were built. The Proton complex benefitted from new assembly and clean rooms. Pads 1 and 31 were renovated. The Energiya-Buran facilities and bays were modernised for integration halls for Western commercial payloads and equipment for the International Space Station. The *Yubeleniye* runway was repaired. The collapse of the Soviet Union presented not only practical problems, but political and legal ones. Baikonur was now in the independent Republic of Kazakhstan. Years of wrangling followed, including rows about landing charges and the cleaning up of chemicals spilled by rockets falling downrange. The outcome was that Baikonur became sovereign Russian territory inside Kazakhstan, paid for through a long-term lease. The Russians considered the lease expensive, but depended on Baikonur for the Proton and Soyuz there. To reduce its vulnerability, Russia first decided to move all military launches to Plesetsk (qv), which continued to

have a strong presence of soldiers. By contrast, Baikonur was civilianised, with all but a small number of soldiers withdrawn in 2006. Secondly, Russia began to give serious attention to constructing a new cosmodrome on Russian territory. This opened as Vostochny (qv) at the same latitude as Baikonur, but in the far east, with a Soyuz pad (2016), to be joined by a pad for the rocket that will replace Proton, the Angara, in 2021. Time will tell if the Russians will eventually leave Baikonur.

Baikonur has become best known to outsiders for Soyuz launches to the International Space Station since 2000. It remains a place of contrasts, some parts being successfully modernised (e.g., pad 31, the Proton area, the assembly and integration halls), but other parts being neglected. It is a target for dark tourists who dare to explore its ruins and post about their exploits online. Markets thrive in the inhabited parts of the city. To visitors, perhaps the most remarkable feature of Baikonur is its ability to conduct its business in all weathers and its clockwork efficiency – both the outcomes of years of practice and refinement. The last time a piloted Soyuz launch was delayed once the countdown had started was 1971. Delays because of heat or, more likely cold, are unknown. Rain and wind make little difference, too, with one cosmonaut recalling that he could feel the cabin rocking in the wind on the pad before take-off. When a group of Indian scientists travelled all the way to Baikonur to see the Russians launch their satellite, thick fog surrounded the site. They presumed that the launch would be called off and prepared to leave. Later it was explained to them that the dim orange glow that they thought they saw in the far distance was actually their satellite taking off.

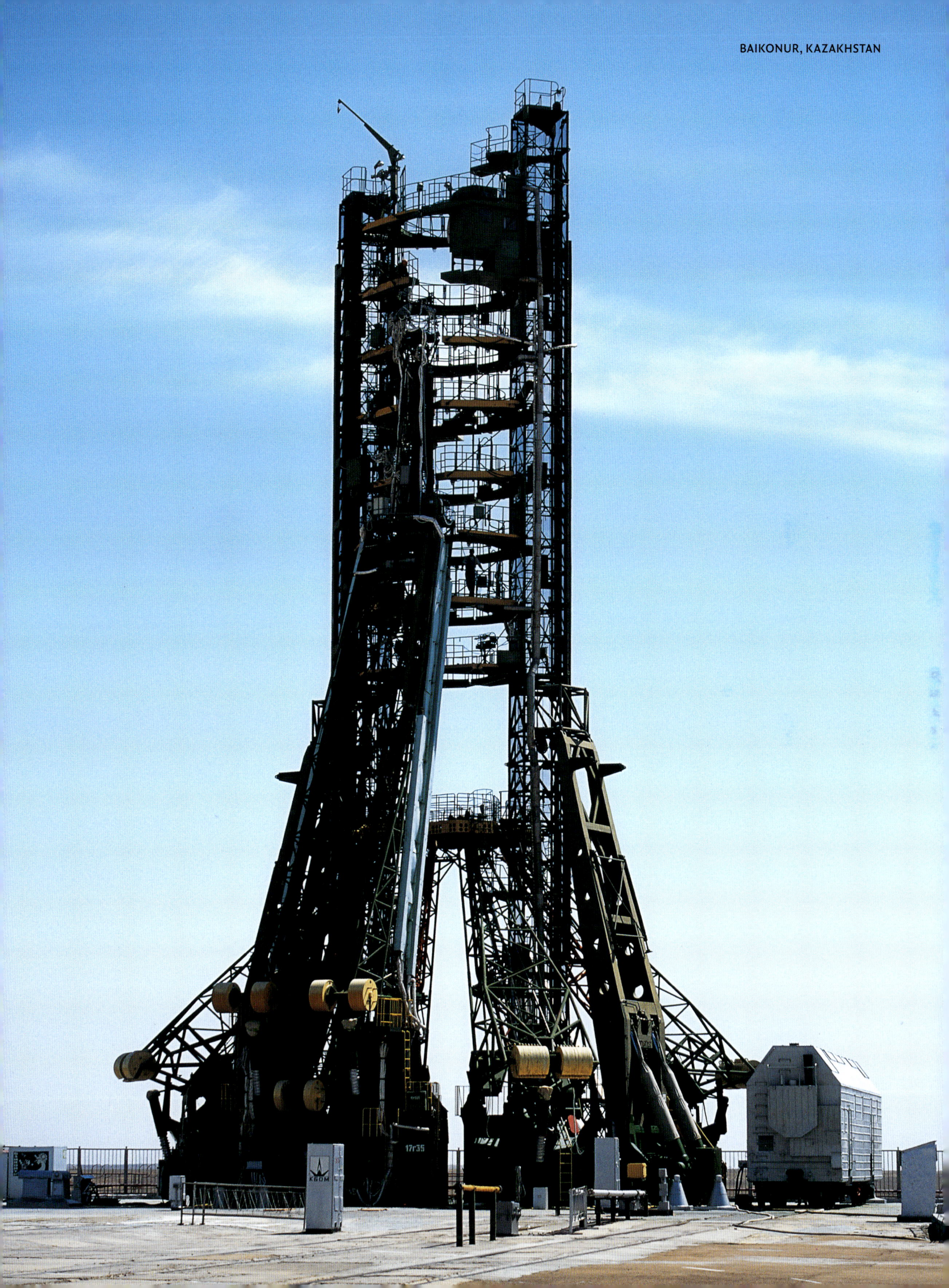

BAIKONUR, KAZAKHSTAN

The large steel launch towers create
a bizarre skyline against the back-
drop of the Kazakh steppe.
Source: DOM publishers Archive

Workers' housing, constructed to not only to provide shelter but also in an attempt to bring normality and familiarity to the interstellar life in the dry desert.
Source: DOM publishers Archive

Unlike the distinct R-7 rockets, the Russian Kosmos follows a more conventional rocket design.
Source: DOM publishers Archive

The Proton rocket family dates back to 1965, offering a reliable and heavy duty launch platform up to the present day.
Source: DOM publishers Archive

Right:
R-7s launched from Baikonur brought humanity into Earth's orbit. These rockets still represent the backbone of not just Russian but international space logistics.

Pages 148–149:
The burnt underbelly of the concrete rocket exhaust structure underneath the Soyuz launch tower.

Pages 150–151:
The enormous flame deflection structure shows its age and use.

Pages 152–153:
Larger than the platform itself, the flame deflection structure lined with concrete panels expands out into the Kazakh steppe.

Source: DOM publishers Archive

Top:
A satellite image of Baikonur city, a Russian-controlled
territory amid the steppes of Kazakhstan (2012).
Source: Google Earth

Right:
A diagram in the Baikonur Museum of History locating
the launch sites north of the settlement (2001).
Source: DOM publishers Archive

INDO-AFRICAN

Baikonur
⊙

Plesetsk
⊙

Kapustin Yar
⊙

Caspian Sea

Sem

Baltic Sea

Black Sea

Palmachim
⊙

Mediterranean Sea

Peenemünde
⊙

Hammaguir
⊙

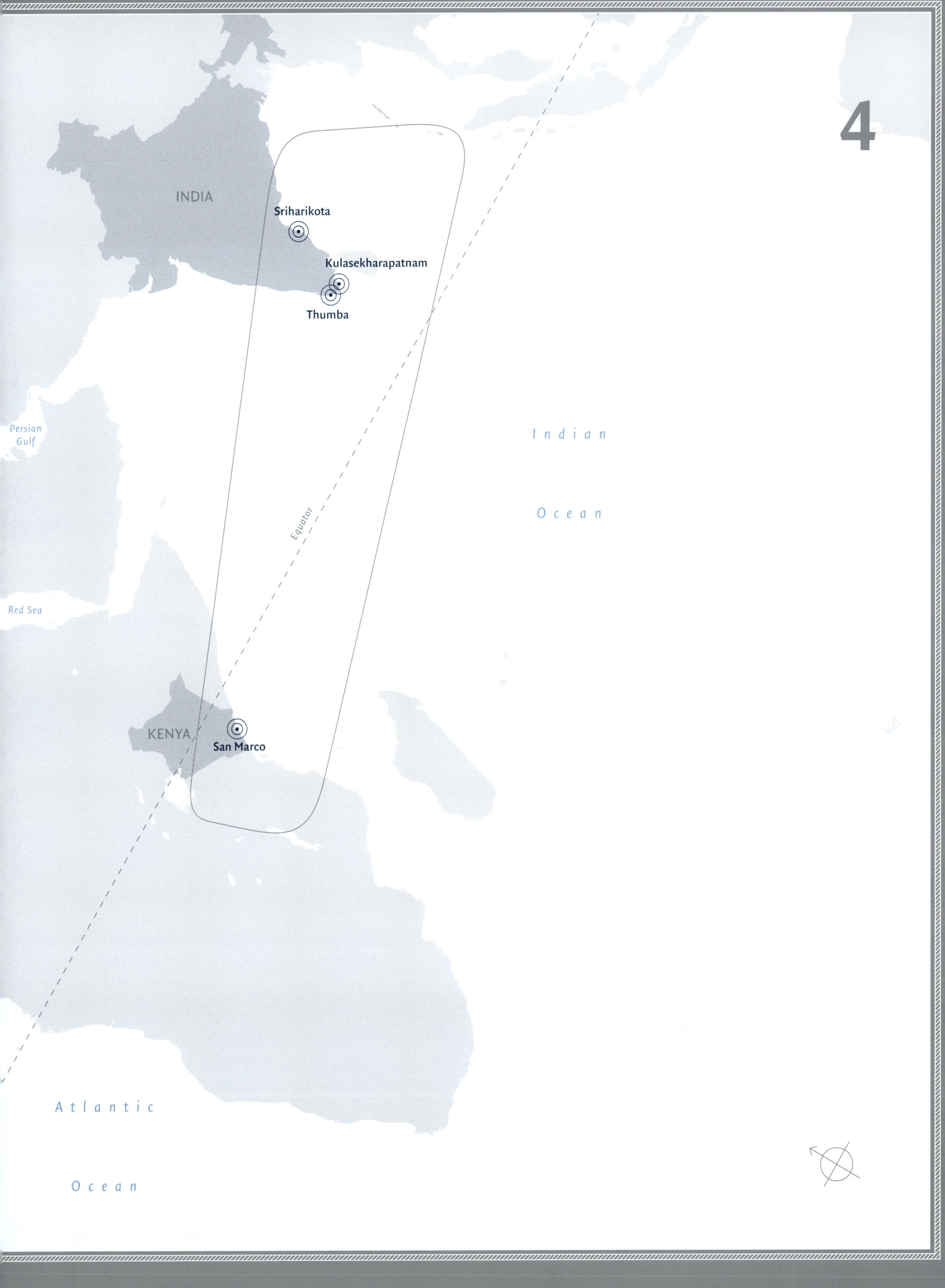

4

INDIA

Sriharikota
⊙

Kulasekharapatnam
⊙
⊙
Thumba

Persian
Gulf

Red Sea

Indian

Ocean

Equator

KENYA
⊙
San Marco

Atlantic

Ocean

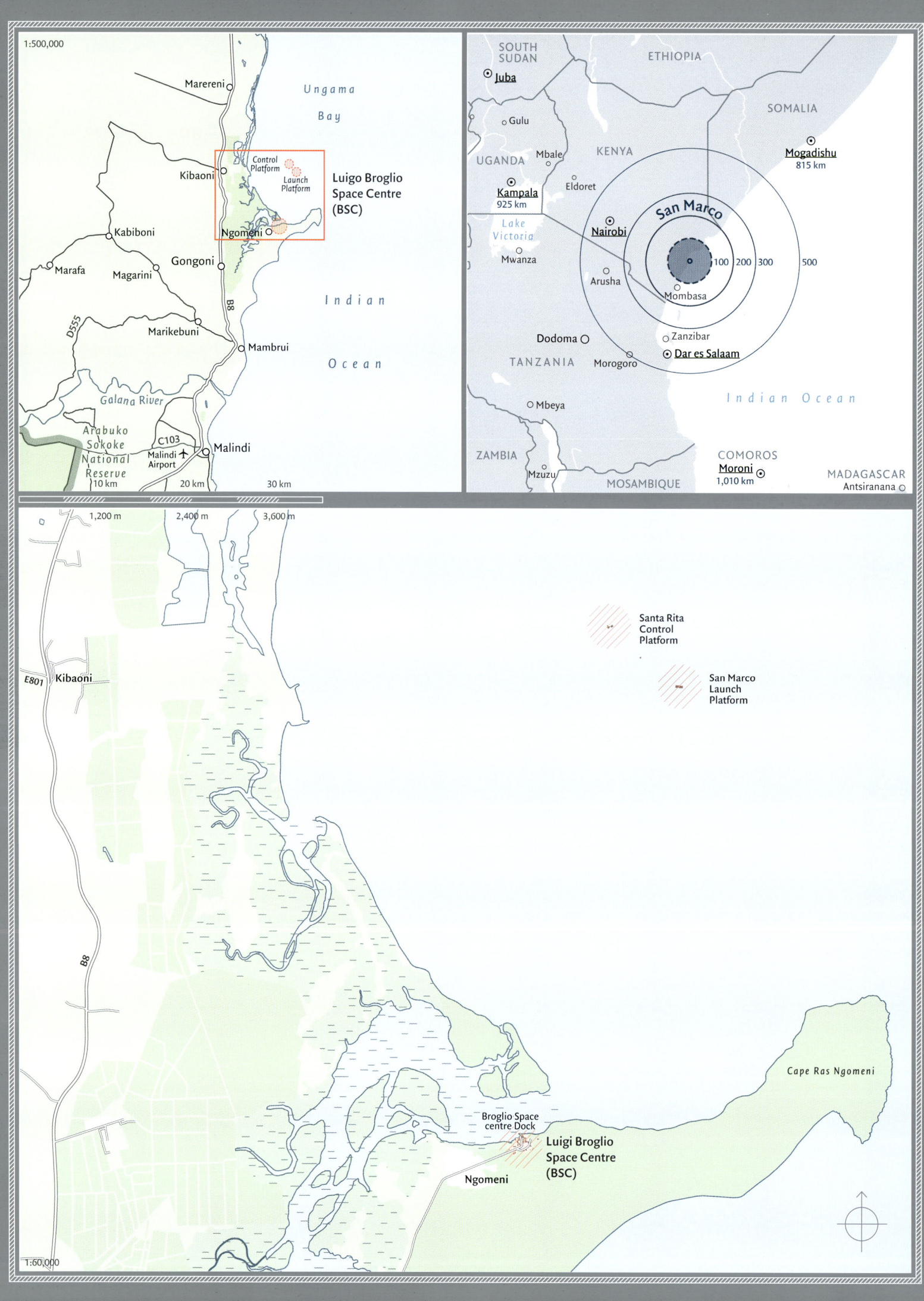

Inset map 1 (top left) — 1:500,000

Marereni

Ungama Bay

Kibaoni

Control Platform

Launch Platform

Luigo Broglio Space Centre (BSC)

Kabiboni

Ngomeni

Gongoni

Marafa

Magarini

Indian

D555

Marikebuni

Mambrui

Ocean

Galana River

Arabuko Sokoke National Reserve

C103

Malindi Airport

Malindi

B8

10 km   20 km   30 km

Inset map 2 (top right)

SOUTH SUDAN

Juba

ETHIOPIA

Gulu

SOMALIA

UGANDA

Mbale

KENYA

Mogadishu
815 km

Kampala
925 km

Eldoret

San Marco

Nairobi

Lake Victoria

100 200 300   500

Mwanza

Arusha

Mombasa

Dodoma

Zanzibar

TANZANIA

Morogoro

Dar es Salaam

Indian Ocean

Mbeya

ZAMBIA

COMOROS

MADAGASCAR

Mzuzu

Moroni
1,010 km

MOSAMBIQUE

Antsiranana

Main map (bottom) — 1:60,000

1,200 m   2,400 m   3,600 m

E801   Kibaoni

Santa Rita Control Platform

San Marco Launch Platform

B8

Cape Ras Ngomeni

Broglio Space centre Dock

Luigi Broglio Space Centre (BSC)

Ngomeni

PIONEERING OFFSHORE LAUNCH PLATFORM

Name		**Coordinates**
Broglio Space Centre		2.9ºS, 40º 12.2ºE
Location		**Time zone**
Kenya, Africa		UTC+06:00
Owner / Operator		**Launches**
Italian Space Agency		9 satellites
Elevation		**Completion**
10 m		1964

San Marco was an early, historic launch site – an Italian, American, and Kenyan project demonstrating lateral thinking in how to launch satellites. The founder of the Italian space programme, Professor Luigi Broglio, developed the idea of converting an oil rig, locating it in international waters, and using the American Scout rocket to launch small Italian satellites from it. With the agreement of the Kenyan government, a site was agreed 5 km off the coast in Ugwana Bay, offering direct ascent to equatorial orbit and a downrange path over the ocean and away from inhabited areas. The launch site, San Marco, was a triangular-shaped platform, towed there from La Spezia, Italy, and embedded in the ocean floor. Boats transported people and equipment from shore. The Scout rocket was brought alongside, lifted aboard, attached to a railing, pointed, and fired. Shore and tracking facilities were set up in Ras Ngomeni and Malindi. Eighty staff worked on the site during launches. The site included a second platform, Santa Rita – the cable-linked command ship set on stilts 500 m away and a support ship. During the three years of construction, Italian engineers went to the United States to train on the Scout. The Scout rocket fired into orbit an Italian scientific satellite, San Marco 2, in April 1967, proving the system. Subsequent launches were San Marco 3, 4, and 5; the famous British Ariel 5; and four American Explorer satellites (42, 45, 48, and 53), including the spectacularly successful first X-ray astronomy satellite, Explorer 42 Uhuru ('freedom', to honour Kenya's independence). The Scout eventually went out of production and the San Marco platform was last used in 1988. There was some political unease that Kenya got little from the project. The site is still there but

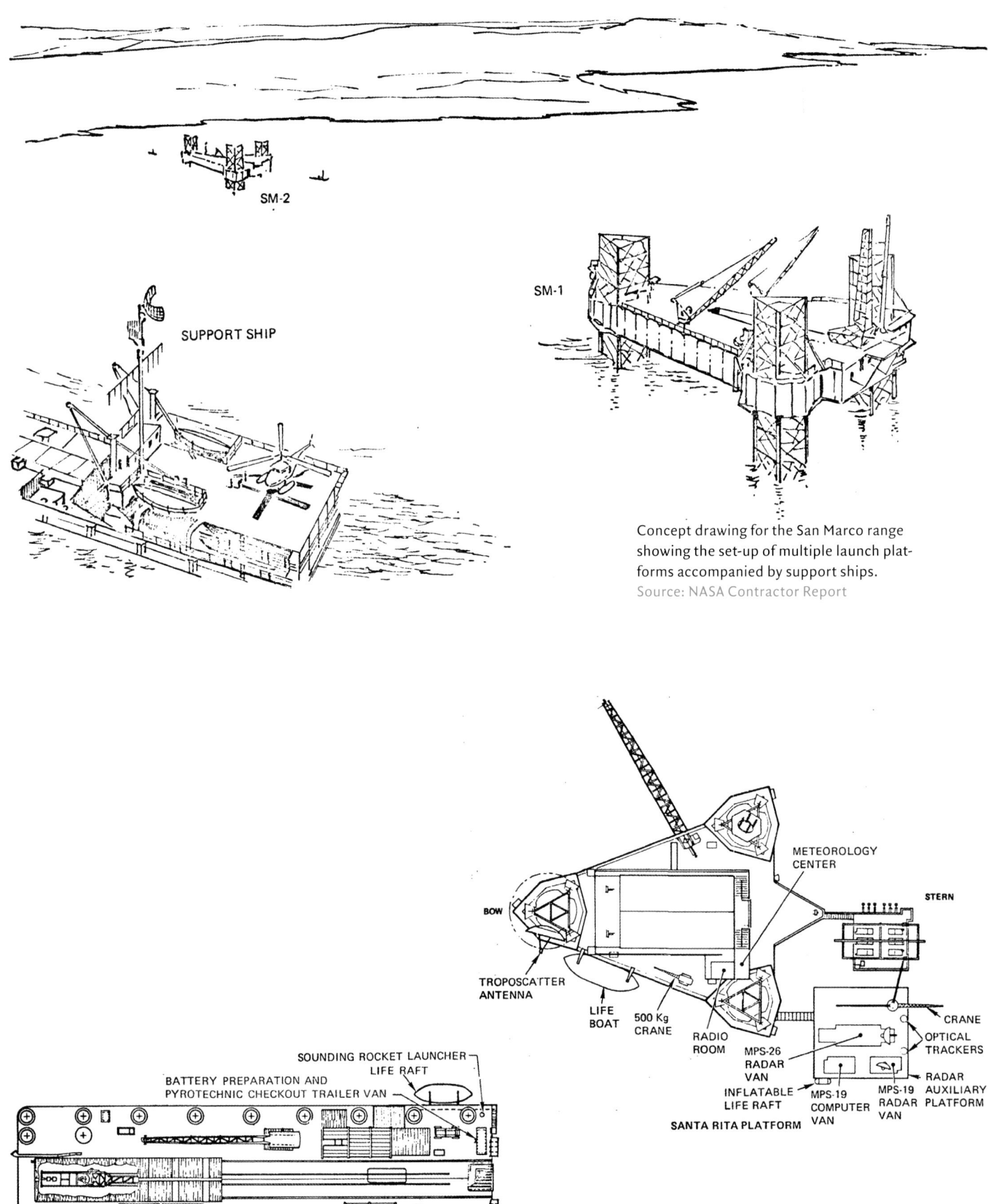

SM-2

SM-1

SUPPORT SHIP

Concept drawing for the San Marco range
showing the set-up of multiple launch platforms accompanied by support ships.
Source: NASA Contractor Report

METEOROLOGY
CENTER

STERN

BOW

TROPOSCATTER
ANTENNA

LIFE
BOAT

500 Kg
CRANE

RADIO
ROOM

MPS-26
RADAR
VAN

INFLATABLE
LIFE RAFT

MPS-19
COMPUTER
VAN

MPS-19
RADAR
VAN

CRANE

OPTICAL
TRACKERS

RADAR
AUXILIARY
PLATFORM

SANTA RITA PLATFORM

SOUNDING ROCKET LAUNCHER
LIFE RAFT

BATTERY PREPARATION AND
PYROTECHNIC CHECKOUT TRAILER VAN

500 Kg
CRANE

CLEAN
ROOM

WORKSHOP
TRAILER VAN

SAN MARCO PLATFORM

Draft drawings for additional improvements to
the two steel rigs of San Marco and Santa Rita.
Source: NASA

Right:
The Ariel 5 launching from San Marco platform in 1974. During its operation, the platform served as a place for joint space efforts between the Italian Space Commission (CRS) and NASA.
Source: NASA

Bottom:
The rectangular launch platform San Marco.

Santa Rita, a refurbished oil platform that housed the main control centre.

Support ship SS Tortugas transferring a rocket and equipment onto San Marco.
Source: NASA

rusting, with occasional attempts by Italy and the Kenyan government to resurrect its use. The platform verified the viability of such a system and inspired the later Sea Launch project (qv). A second legacy is that Malindi, still staffed by Italians and leased from the Kenyan government, continued as a tracking and data station for American and European satellites, later becoming an important station for the Chinese space programme. In 2003, the Italian government renamed the system the Luigi Broglio Space Centre and it is still legally part of the Italian Space Agency (ASI).

Aerial photograph of San Marco rocket launch site off the Kenyan coast near Malindi, taken on 14 March 2018.

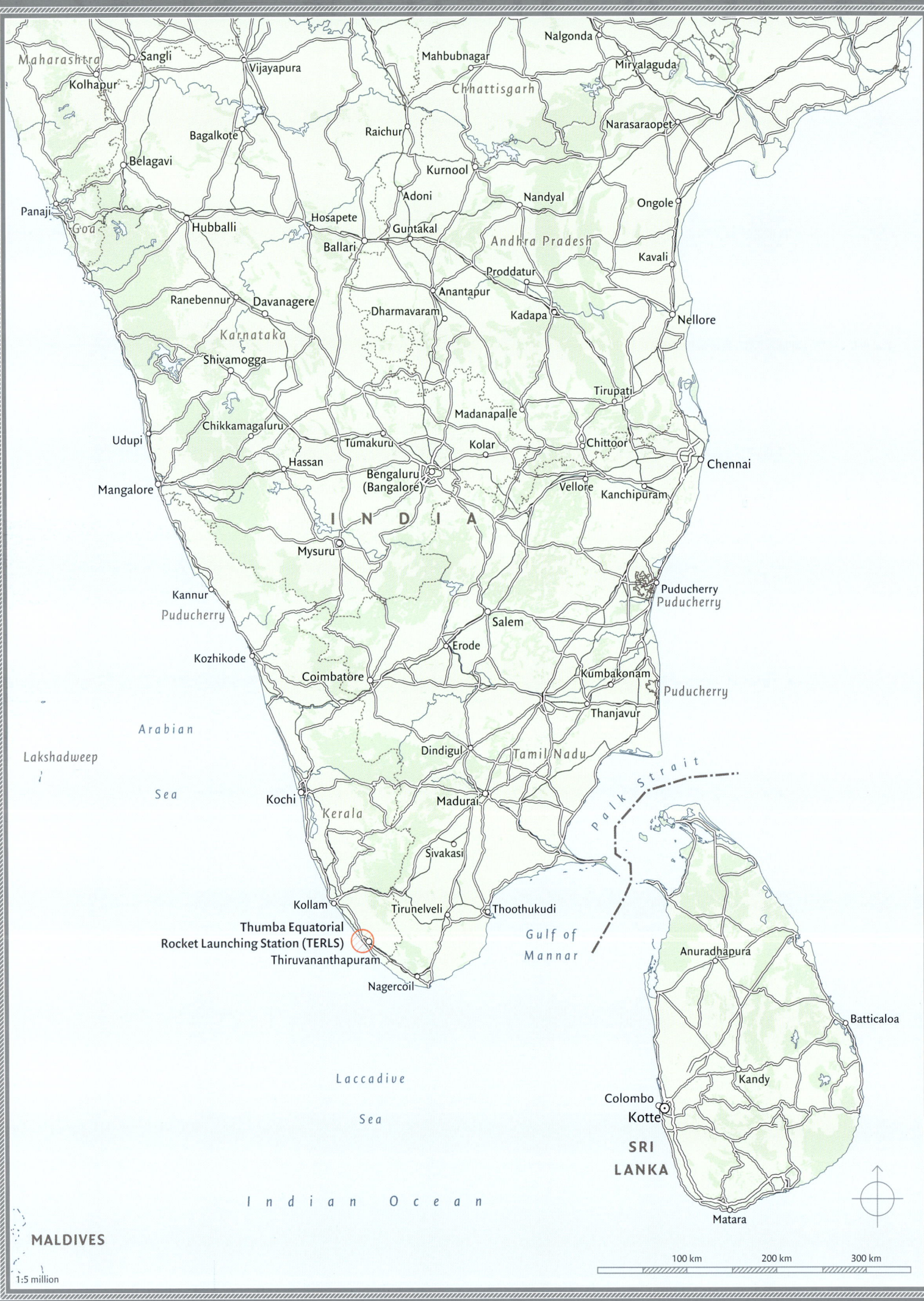

CRADLE OF INDIA'S SPACE PROGRAMME

Name
Thumba Equatorial Rocket
Launching Station

Location
Thiruvananthapuram,
India

Owner/Operator
Indian Space Research
Organisation

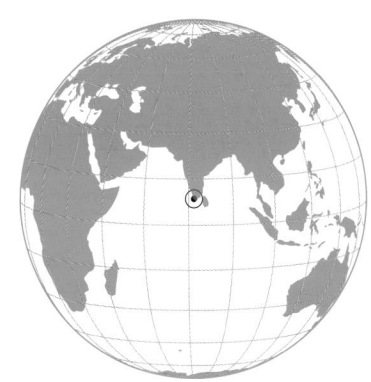

Elevation
6 m

Coordinates
8.5°N, 76.8°E

Time zone
GMT+5:30

Completion
1963

As the sun set on Thursday 21 November 1963 over the coast of the southern Indian state of Kerala, many residents saw something they had never seen before: a glowing orange streak rising high in the sky over the Arabian Sea. What they saw was an expanding cloud of sodium as it escaped from a cylinder while climbing through an altitude of 100 km and crossing into space. This was the first object to reach space launched from Indian soil and it marks the birth of India's space programme. The rocket carrying the sodium payload was launched from the Thumba Equatorial Rocket Launching Station, a rocket launch site that was built in haste and given the political nod just a year earlier.

Named after a plant that grew there in abundance, Thumba is located 14 km from the state capital Trivandrum (now Thiruvananthapuram). It is a parcel of land about 2.5 km long with a railway line to the east and the Arabian Sea to the west and just 8.5°N of the equator. Thumba was not selected for its ecology, but rather its geography. Discovered in the 1920s, the equatorial electrojet (EEJ), a flow of charged particles created by the solar wind, circles the Earth. Thumba is one of a few places on Earth from where rockets can investigate the EEJ 100 km directly above. The first scientific investigation of the EEJ was conducted using sounding rockets launched from Thumba.

The establishment of Thumba required the resettlement of a small fishing community to a new location in the nearby village of Pallithura. The vacated St. Mary Magdalene Church became the centre of the Thumba Space Centre. Abdul Kalam, who subsequently rose to high political office as the President of India, was responsible for integrating that very first payload onto the launch

The nose cone of a Centaur rocket being transported 2 km to the launch pad by rocket engineer C. R. Sathya and Velappan Nair (with the bicycle). French photographer Henri Cartier-Bresson took this photograph in February 1969.
Source: ISRO

vehicle inside this church. He recalled later this unlikely unifier of India's diversity: Kalam (a Muslim), Sarabhai (a Hindu), and a Christian church. Sounding rockets from many countries including the USA, Russia, France, Japan, and Britain have been launched from Thumba. They continue to be launched from Thumba today, albeit with a much-reduced frequency. Following the premature death of Vikram Sarabhai in December 1971, Thumba was consolidated with nearby centres including the original church building, now converted into a museum, and the Vikram Sarabhai Space Centre (VSSC) was established. The VSSC is the largest of ISRO's many centres spread across India.

Vikram Sarabhai developed a network of contacts in Europe, the US, and the USSR and cultivated Thumba to become an international facility for investigating the upper atmosphere as well as the EEJ. On 2 February 1968, in the presence of several international dignitaries, Thumba was formally dedicated as an international range for scientific research and open to all UN member states. It is through this role during the early days of the Cold War that scientists and engineers from the USSR, Japan, the US, and Europe were able to meet and collaborate on science in an apolitical environment that could only exist outside the US, the USSR, and Europe.

INDIA'S NEW LAUNCH SITE

Name		**Coordinates**
Kulasekarapattinam		8.2°N, 78.3°E
Location		**Time zone**
Tamil Nadu, India		UTC+5:30
Owner / Operator		**Launches**
Indian Space Research Organisation		N/A
Elevation		**Completion**
4 m		Expected around 2024

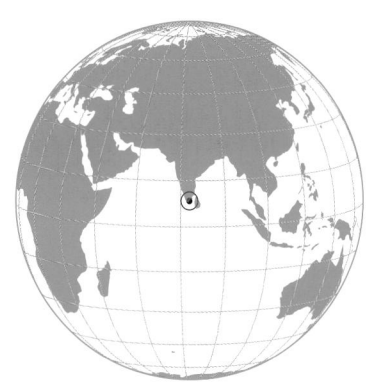

On 6 February 2020, the Minister of State confirmed in the Indian parliament that India would establish another launch site in the Toothukudi district in Tamil Nadu. Many candidate sites in southern India had been in consideration for several years. This decision was the culmination of a debate that has been ongoing for almost a decade. As welcomed as it will be, this will be India's second site capable of orbital launches. Currently, the US has four, Russia five, China four, and Japan has had two since the 1970s.

The site consists of an area measuring 951 ha near Kulasekarapattinam. It is currently occupied by three villages (Mathavankurichi, Padukapathu, and Pallakurichi). Reminiscent of the days when Thumba was established in 1963, re-location of the community is estimated to affect 27 families. All the families impacted will be offered compensation, land, and new homes. The local community in Tamil Nadu is expected to benefit from around 10,000 new jobs resulting directly and indirectly from this initiative. The site is located about 648 km south of Chennai. The location has two key advantages: its proximity to ISRO's Liquid Propulsion Systems Centre at Mahendragiri, which is just 70 km away, and its location relative to Sriharikota. It is further south and west. Launching due south for a polar orbit will not require a trajectory change to avoid Sri Lanka. An easterly launch for an equatorial orbit will also require a reduced trajectory manoeuvre to avoid flying over Malaysia. In both cases, the fuel saved will increase the payload capacity.

Kulasekarapattinam will be the primary site for India's new Small Satellite Launch Vehicle (SSLV), although the initial three development flights of the SSLV will be conducted at Sriharikota before

Puffs of smoke and the deafening roar from an engine test conducted at the ISRO Propulsion Complex in Mahendragiri. Soon this cacophony will be heard across Tamil Nadu's beaches.
Source: ISRO

the SSLV is deemed operational. The SSLV is a three-stage all solid-fuel launch vehicle measuring 34 m in height and 2 m in diameter, with a lift-off mass of 120 tonnes. It can deliver 500 kg to equatorial orbit or 300 kg to a polar orbit of about 500 km.

ISRO has established a commercial arm called New Space India Limited that will engage companies in India's and the growing international private space sector to take advantage of the increasing demand for small satellite launch capacity. The SSLV has a tentative payload fairing of 2.9 m with a diameter of 1.9 m. The initial SSLV marketing materials show that 2, 24, or 64 satellites of varying sizes can be launched at one time.

An unambitious launch cadence for the SSLV is advertised to be between six and eight launches per year. In addition to a traditional fixed launch pad, ISRO has indicated the possibility of mobile launchers. The 951 ha at Kulasekarapattinam has extensive potential to facilitate a mobile launch service. The SSLV concept was motivated by the increasing global demand for the launch of small satellites. The initial operational date was intended to be in 2021. Delays arising from the Covid-19 pandemic pushed the first launch to August 2022.

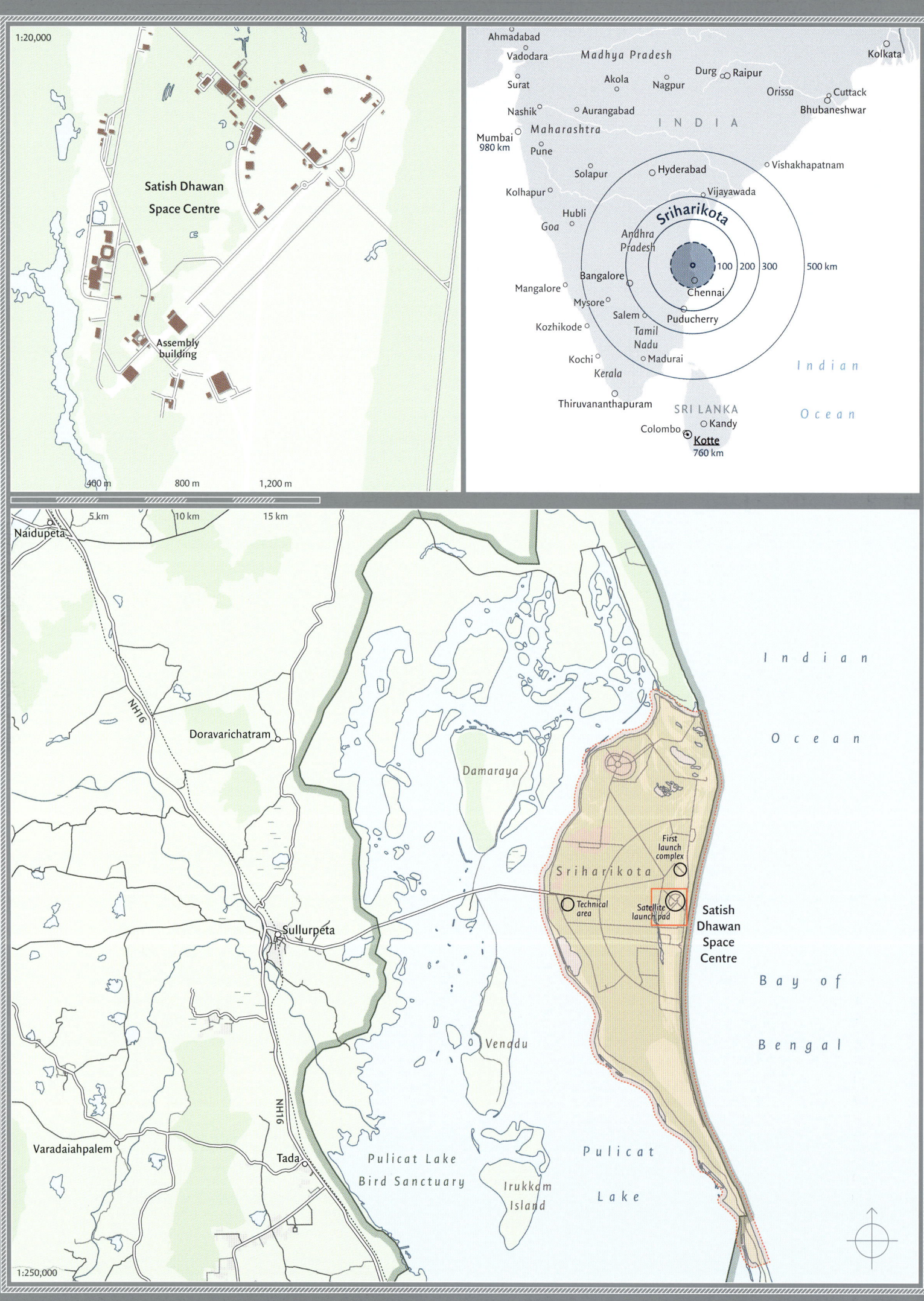

1:20,000

Satish Dhawan
Space Centre

Assembly
building

400 m   800 m   1,200 m

Ahmadabad
Vadodara   *Madhya Pradesh*
Surat   Akola   Nagpur   Durg   Raipur   Kolkata
Nashik   *Aurangabad*   *Orissa*   Cuttack
Maharashtra   Bhubaneshwar
Mumbai
980 km   Pune
Solapur   Hyderabad   Vishakhapatnam
Kolhapur   Vijayawada
Hubli   *Andhra*   **Sriharikota**
Goa   *Pradesh*
100 200 300   500 km
Bangalore
Mangalore   Mysore   Chennai
Salem   Puducherry
Kozhikode   *Tamil*
Nadu   *Indian*
Kochi   Madurai
Kerala   *Ocean*
Thiruvananthapuram
SRI LANKA
Colombo   Kandy
Kotte
760 km

I n d i a n

O c e a n

Naidupeta
5 km   10 km   15 km

NH16
Doravarichatram

Damaraya

First
launch
complex

Sriharikota

Technical
area   Satellite
launch pad

Satish
Dhawan
Space
Centre

Sullurpeta

B a y   o f

B e n g a l

Venadu

NH16

Varadaiahpalem
Tada   *Pulicat Lake*
Bird Sanctuary   *P u l i c a t*

L a k e
Irukkam
Island

1:250,000

INDIA'S GATEWAY TO SPACE

Name	Coordinates	
Satish Dhawan Space Centre	13.7°N, 80.2° E	

Location	Time zone
Nellore District, India	GMT+5:30

Owner/Operator	Launches
Indian Space Research Organisation	537 satellites

Elevation	Completion
2 m	1979

All the spacecraft launched from India to Earth orbit and beyond started their journey from Satish Dhawan Space Centre, also known as Sriharikota. Most of India's around 50 operational spacecraft today were launched from Sriharikota, including both missions to the Moon and one to Mars. In 2018, the Indian government tasked the Indian Space Research Organisation (ISRO) with commencing its programme to take Indian astronauts to Earth orbit using an Indian launch vehicle. Originally scheduled for 2022, whenever that happens, Indian astronauts will embark on their journey to space from Sriharikota.

Sriharikota covers an area of 145 km². It was first identified in 1968 and became operational in October 1971, a few months before the unexpected demise of Vikram Sarabhai, who is regarded as the founder of the Indian space programme. Similar to Cape Canaveral on Merritt Island in the US, Sriharikota is on a barrier island – a stretch of land 8 km wide sandwiched between Buckingham Canal and Pulicat Lake. It is on the east coast about 100 km north of Chennai. Pulicat Lake is a protected area with rare and beautiful wildlife including kingfishers, pelicans, and pink flamingos.

The Yanadi community that had lived in the area and was suppressed repeatedly for centuries was forcibly removed in 1969 to make way for the launch site. Like many other launch sites established in remote locations around the world, the displaced indigenous communities first lose their homes and then rarely benefit from the technology designed to extricate communities like theirs from poverty.

Satish Dhawan (right) stepped in to lead ISRO following the sudden death of Vikram Sarabhai in 1971. A. P. J. Abdul Kalam (left), Project Director for the SLV-3, India's first launch vehicle to place a payload in Earth orbit in July 1980.
Source: ISRO

Right:
Geosynchronous Satellite Launch Vehicle Mk II, in operation since 2010. It is a three-stage launch vehicle capable of using solid, liquid, and cryogenic propellent. It is capable of placing a payload of 2.3 tonnes in Geo Transfer Orbit.
Source: ISRO

Sriharikota became operational in 1971 with the launch of a sounding rocket, but it was a decade later that the first successful orbital launch was conducted using India's Satellite Launch Vehicle (SLV-3). Half a century on, Sriharikota remains India's only operational launch site to Earth orbit. All of India's launch vehicles (SLV-3, ASLV, PSLV, and the GSLV) have been developed and launched from Sriharikota. Sounding rockets have also been launched from Sriharikota and that continues today. It is also used to test experimental technologies including scramjet and reusable launch vehicles. Between 1979 and 2021, 82 orbital launches have been conducted from Sriharikota, though not all have been successful.

The original launch pad built in 1979 was used for the SLV-3 and Augmented Satellite Launch Vehicle (ASLV) and used horizontal integration. That launch pad was decommissioned in 1994. Today Sriharikota has two launch pads: the First Launch Pad (FLP), operational in 1993, and the Second Launch Pad (SLP), operational in 2005. Both launch pads use vertical integration. Launch operations are supported by two Vehicle Assembly Buildings (VAB and the Second VAB or SVAB), a Mission Control Centre (MCC) incorporating two Launch Control Centres (LCC), and a Solid Propellant Booster Plant (SPROB). A visitor complex opened in 2019. It has a seating capacity of 10,000, a space museum, space theatre, and a rocket garden.

Sriharikota is on India's east coast at a latitude of 13°N. Its geography imposes launch constraints. Sri Lanka to the south gets in the way for a polar orbit and Malaysia and Indonesia (including the busy shipping lanes and an offshore oilfield) in the east are in the way for an equatorial orbit. GSLV launch trajectory for an equatorial orbit is not due east at 90° but 108° with an essential trajectory manoeuvre once the launch vehicle is beyond Malaysia. Similarly, a typical PSLV polar orbit trajectory is not due south at 180° but 135°, requiring a trajectory manoeuvre around Sri Lanka.

Since 2010, the FLP has been used exclusively for the launch of the PSLV; it cannot accommodate the larger GSLV-Mk3. The launch vehicle is integrated vertically at the FLP using a 60-tonne crane on top of a 76-metre-high, 3,000-tonne Mobile Service Tower (MST) with four retractable platforms providing access to each of the PSLV's four stages. Prior to launch, the MST is moved 150 m away at 7.5 m per minute on twin rails. A PSLV Integration Facility (PIF) is currently under construction near the FLP. Conceptually, the PIF is another VAB for the FLP. It is 60 m high, 30 m wide, and 35 m long with nine foldable platforms providing access along the length of the PSLV. The 1,050 m² floor space enables work on the PSLV, S139 solid rocket boosters, and the recently introduced Small Satellite Launch Vehicle (SSLV) to proceed simultaneously. Once a launch vehicle integration is complete, a Mobile Launch Platform (MLP) transports it by rail to the FLP. Construction is due to be completed by 2022, and this will transform the mode of operation at the FLP from 'integrate and launch' to 'integrate, transport, and launch' as the approach taken at the SLP. With the support of the private sector, ISRO expects to double the launch cadence of the PSLV to about 15 every year.

The first stage and the two solid boosters for a GSLV are integrated in the Solid Stage Assembly Building (SSAB), which is 58 m high, 55 m long, and 40 m wide, and then transported the short distance by rail to the larger VAB (83 m × 40 m × 32 m). Once the integration

India's first 24-hour orbit communication satellite, APPLE, prior to launch in 1981. A bullock cart with rubber tyres provided a pragmatic and inexpensive non-magnetic environment to conduct antenna tests in an open field.
Source: ISRO

is complete, the launch vehicle is transported 1 km by rail to the SLP on a 16-wheel MLP measuring 19.5 m × 19.5 m and weighing 700 tonnes. Once at the SLP, a 70-metre-tall umbilical tower facilitates remote control, communication, and fuelling to prepare the vehicle for launch. The SLP is supported by an 80-metre-high water storage tower containing 50,000 tonnes of water released in 20 seconds at launch into the exhaust pit to minimise the vibration through acoustic dampening. The SLP continues to be used to launch the PSLV, the GSLV-Mk2, or the GSLV-Mk3.

With its larger capacity, the SVAB can accommodate the larger launch vehicles in ISRO's future plans. The SSAB is 96 m high, 70 m long, and has a width of 50 m. It has a crane with a span of 35.5 m looking down onto a floor space of 3,500 m². Unlike the VAB, the SVAB does not have a SSAB. The GSLV's two large Solid Rocket Boosters are integrated within the SVAB. The SVAB was inaugurated in 2019, but has yet to be fully commissioned. Both launch pads are designed to withstand the annual Indian cyclone season and winds of up to 230 km/hr between September and December. Traditionally, launches are not conducted during this period. India's first mission to Mars launched on 5 November 2013 was an exception to catch the narrow launch window.

All the solid propellent used by ISRO's launchers, primarily the solid stage strap-on boosters for the PSLV and GSLV, are produced at Sriharikota in the SPROB. The facility comprises two separate plants for the manufacture of the Hydroxyl-Terminated Polybutadiene (HTPB) used for the first and third stage of the PSLV and the strap-on boosters for the PSLV and GSLV. The Static Test and Evaluation Complex (STEX) is also in Sriharikota. Every batch of propellent is quality tested at the STEX to provide quantitative measurements of the physical and propulsive characteristics to ensure a payload attains its desired orbital altitude.

Opened by the Indian Prime Minister in 2012, the MCC is 6 km from the launch pads and at the heart of all operations at Sriharikota. A stunning design, the huge building looks like a white flying saucer built on an elevated mound with good visibility of the launch pads. The MCC is the hub for all audio and video communications, telemetry, range operations, meteorology, and live video feed from the launch pad. The MCC contains two identical Launch Control Centres (LCC) for redundancy. Both LCCs are connected to both launch pads and redundancy is supported further by each desk in each LCC having two monitors, an uninterruptible power supply, and back-up diesel generators.

The Sriharikota launch cadence remains stubbornly low. Despite the two launch pads, the maximum number of annual launches has not exceeded seven (in 2016 and 2018). A proposal for a third launch pad was replaced (at least for the present) in favour of the PIF at the FLP. Covid-19 restrictions in early 2020 further delayed launches. However, ISRO's liquid oxygen production facilities were used to support the national shortage of oxygen during the pandemic.

ISRO's heavy-lift launcher, the GSLV-Mk3, was finally operationalised in on its fourth flight in July 2019. Since the maximum payload capacity of the GSLV-Mk3 to GEO is 4 tonnes, ISRO continues to rely on Ariane-5 operated by Arianespace to launch its heavier communication satellites.

A design for a runway at Sriharikota to support the Reusable Launch Vehicle programme has been in place for some time but not yet started. Both launch pads are about 10 m above sea level and only a few hundred metres from the seashore. A repeat of the Indian Ocean tsunami of 2004 could cause major disruption. Driving to Sriharikota involves traversing a single elevated road over the low-level Pulicat Lake on either side. In extreme weather, this could be subject to flooding. Both possibilities have a low probability but a high impact. Given that Sriharikota is India's only launch site, additional facilities elsewhere will offer essential redundancy.

Since 1979, 563 suborbital sounding rockets and 82 orbital launches, that have included 12 student satellites, 2 re-entry missions and launched 342 satellites for 34 countries. Originally known as the Sriharikota High Altitude Range (or SHAR, which means 'arrow' in Sanskrit), it was renamed in 2002 as the Satish Dhawan Space Centre after its director between 1972 and 1984.

Top:
Opened in 2012, ISRO's flying saucer-like Mission Control Centre is 6 km from the launch pads. It is at the heart of all the action on launch days when all support staff vacate the launch pad. All launch pad operations are conducted remotely from here.
Source: Picture Alliance / AP Photo / Arun Sankar K

Right:
Launched in April 2019, PSLV C-45, a new variant PSLV-QL, placed EMISAT (an Indian surveillance satellite) and 28 additional satellites for international customers into three different orbits. The fourth stage was restarted three times before arriving at its final orbit to perform its new role as an orbital platform with three onboard payloads for research for a further six months before re-entry.
Source: ISRO

Left:
GSLV-F10/EOS-03 on its way to the launchpad. The First Vehicle Assembly building on the right, with the Second VAB on the left.
Source: ISRO

Left:
Paramilitary soldiers stand guard near the Polar Satellite
Launch Vehicle (PSLV-C25) at the Satish Dhawan Space Centre
at Sriharikota on 30 October 2013. India's Mars orbiter mission
was launched by the Polar Satellite Launch Vehicle (PSLV-C25)
on 5 November 2013.

Source: Picture Alliance / AP / Arun Sankar K

An undated handout picture provided by the Indian Space Research
Organisation on 22 June 2016 shows the fully integrated PSLV-C34, with all
its 20 satellites, being moved out of the vehicle assemble building to the
second launch pad at Sriharikota's Satish Dhawan Space Centre in Andhra
Pradesh, India. ISRO set a new record after successfully launching 20
satellites to orbit, including 17 foreign satellites, in a single mission from
its Satish Dhawan Space Centre.

Source: Picture Alliance / dpa / ISRO

On 8 Sept 2016, GSLV-F05 placed an advanced weather
satellite, INSAT-3DR into a geo-transfer orbit. This was ISRO's
tenth flight of its Geosynchronous Satellite Launch Vehicle.
Source: Picture Alliance / dpa / ISRO

Bright objects amid an ocean of green leaves:
GSAT-6 was the 25th geostationary communication satellite
of India built by ISRO. It was launched by the Geosynchronous Satellite
Launch Vehicle (GSLV) rocket from Sriharikota on 27 August 2015.
Source: ISRO

Top: PSLV-C42 carrier rocket on the First Launch Pad with the Second Launch Pad in the background, 16 September 2018.
Source: ISRO

Right: GSLV-F10 with the EOS-3 satellite at the launch pad on 12 August 2021. The mission was unsuccessful due to a technical anomaly in the cryogenic stage.
Source: Picture Alliance / Javed Dar

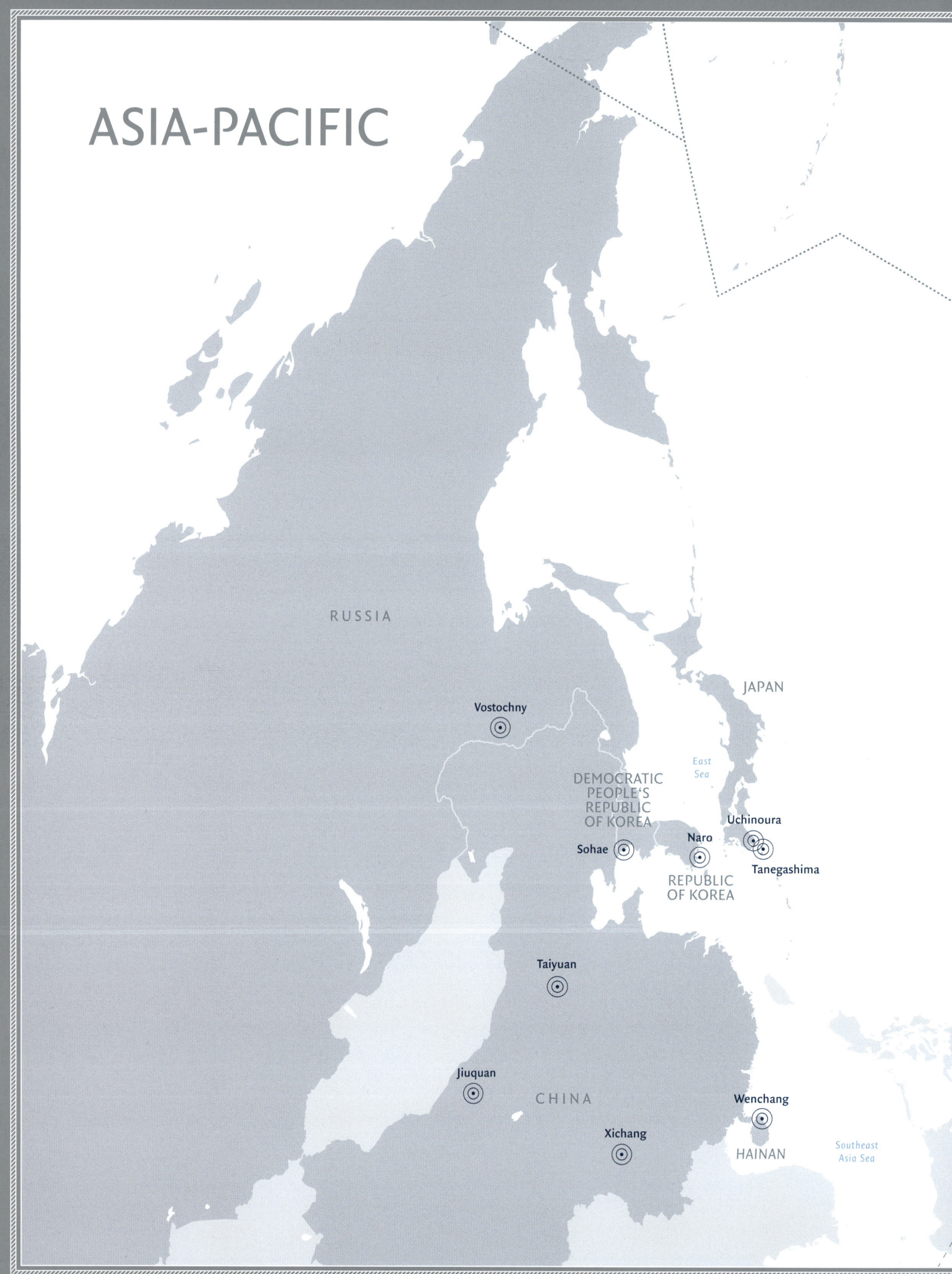

ASIA-PACIFIC

RUSSIA

Vostochny

JAPAN

DEMOCRATIC
PEOPLE'S
REPUBLIC
OF KOREA

East
Sea

Uchinoura

Naro

Sohae

Tanegashima

REPUBLIC
OF KOREA

Taiyuan

Jiuquan

CHINA

Wenchang

Xichang

HAINAN

Southeast
Asia Sea

Sea Launch

Mahia

Pacific

Ocean

Equator

NEW
ZEALAND

AUSTRALIA

Woomera

*Great
Australian
Bight*

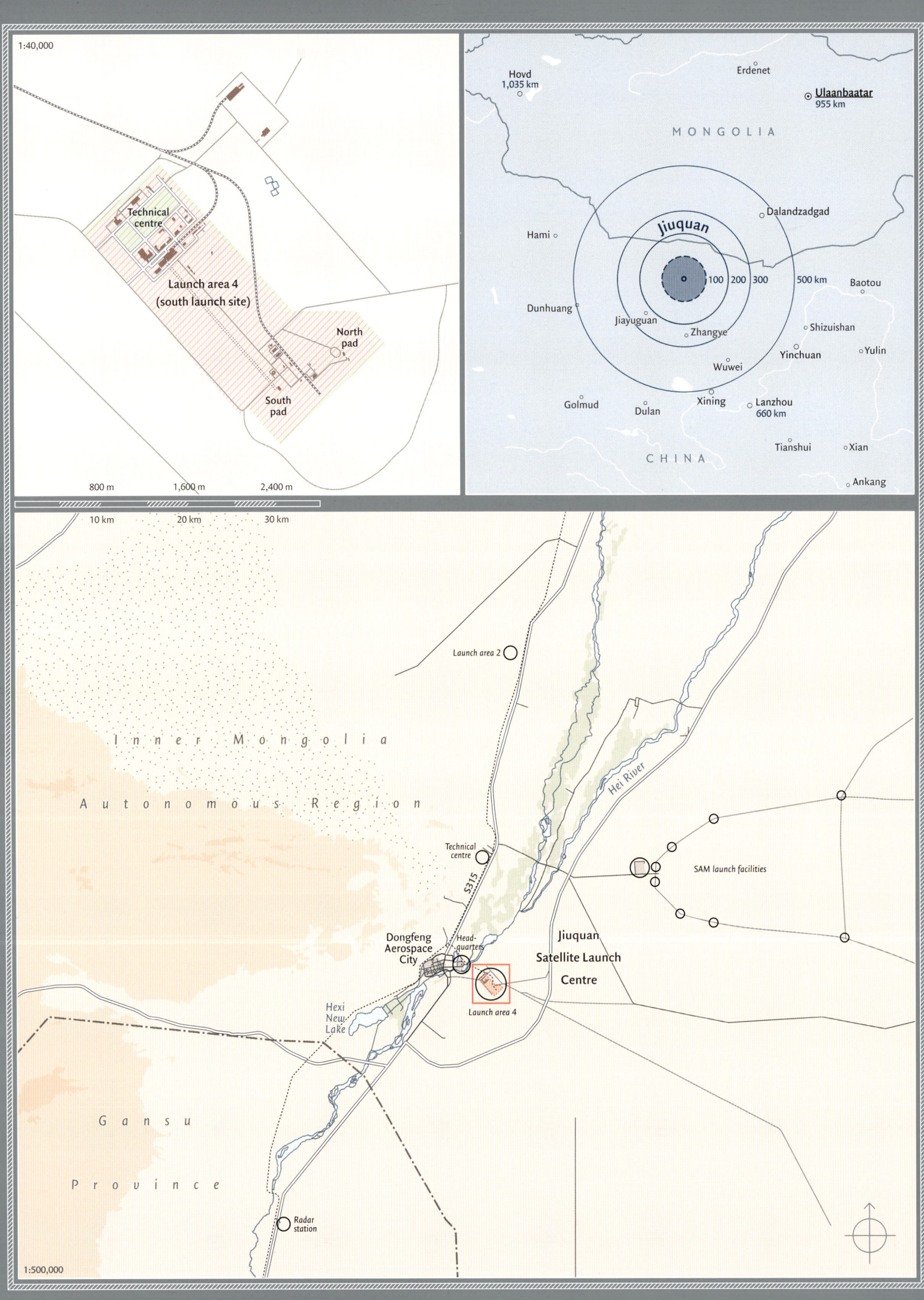

1:40,000

Technical centre

Launch area 4
(south launch site)

North pad

South pad

800 m   1,600 m   2,400 m

Hovd
1,035 km

Erdenet

Ulaanbaatar
955 km

M O N G O L I A

Dalandzadgad

Hami

Jiuquan

100   200   300   500 km

Baotou

Dunhuang

Jiayuguan   Zhangye

Shizuishan

Yinchuan

Yulin

Wuwei

Golmud   Dulan   Xining   Lanzhou
660 km

C H I N A

Tianshui   Xian

Ankang

10 km   20 km   30 km

Launch area 2

I n n e r   M o n g o l i a

A u t o n o m o u s   R e g i o n

Hei River

Technical
centre

S315

SAM launch facilities

Dongfeng
Aerospace
City

Head-
quarters

Jiuquan
Satellite Launch
Centre

Launch area 4

Hexi
New
Lake

G a n s u

P r o v i n c e

Radar
station

1:500,000

CHINA'S PREMIER LAUNCH SITE

Name
Jiuquan Satellite
Launch Centre

Location
Ejin
Inner Mongolia

Owner / Operator
China National Space
Administration

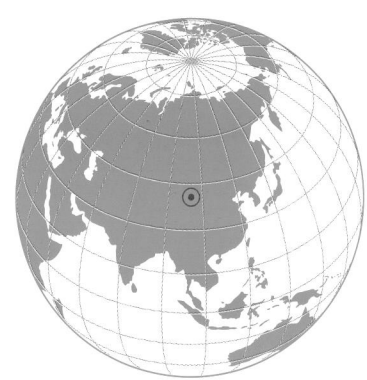

Elevation
1073 m

Coordinates
40.6°N, 99.9°E

Time zone
GMT+8

Launches
273 satellites

Completion
1970

Chairman Mao Zedong decided in 1958 that China should build an orbital satellite, so a distant launch site was located in the Gobi Desert in north-west China. It was given the military title of 'Base 20', guarded by the People's Liberation Army, and equipped with surface to air missiles. Despite being far inland, it was soon found by American U-2 spy planes operating out of Japan. Jiuquan was built during the time of the China-USSR alliance (1954 – 1960) and followed Soviet design and architectural features. It was a rail-based system, with rockets hauled there on a two-day journey all the way from Beijing. The first launch site, though, was a simple set of two 60 m by 60 m concrete pads, north and south, 300 m apart, with a shared 11-floor steel tower used to home the Long March CZ-1 rocket, brought there by a 55-metre-tall, 1,400-tonne shared moveable service gantry running on 17-metre-wide rails. Rockets

were first assembled on the ground in a horizontal position, towed to the pad by tractor, and then reassembled vertically on the pad by cranes. The tower had supporting arms to provide fuel, gas, and electricity right up to the final moments. Mission control used the Russian system of an underground control bunker with a periscope 200 m away. The main processing building was 140 m long with an area of 4,587 m². It contained a 90 m by 18 m test and assembly hall and a 24 m by 8 m fuelling hall where equipment could be moved around by a crane capable of lifting 16 tonnes. Adjacent were 25 test rooms for checking out spacecraft parts. Beside them was a solid rocket motor checkout and processing hall measuring 24 m by 12 m, with a crane and storage and test facilities. The halls guaranteed clean room standards of 100,000 class (one dust particle in 100,000 or less), temperatures of 20.5°C, and humidity in

Snow lies on the now-frozen steppe as a CZ-2C
lifts off in November 2009.
Source: Picture Alliance / dpa / Maxppp Xu Haihan

Rockets launched here must be able to resist both the
dry and hot desert climate and the cold.
Source: Picture Alliance / dpa / Maxppp Chinafotopress

Left:
The distinct blue launch pad that hosted the rockets
used in China's mission to build its space station.
Source: Picture Alliance / dpa / Js Ic Wh

the 35 to 55 per cent range. Fuels were stored in underground bunkers. There were four-storey apartment blocks for the workforce. China's space programme was delayed by the Great Leap Forward and then the Cultural Revolution, but its first satellite was finally launched from here on 24 April 1970. This set of launch pads was used for its successor, the Long March CZ-2, the space programme having a relatively slow launch rate over its first 30 years. Jiuquan launches curved over to the south-east passing over an observation site 4 km to the east. Conditions in the early years were quite basic: surface transport was by military trucks and by rail on steam trains. Visitors watched launches from collapsible chairs on the sand, with loudspeakers on telephone poles to relay the countdown and commands given to personnel by whistles and flags.

The nearest town, 70 km away, was the city of Jiuquan (meaning 'oasis'). It marks the end of the forts of the Great Wall, was part of one of the old Silk Roads, and is now a tourist destination. It is a modern, well-equipped, prosperous city, laid out in a grid, and softened by wind-breaks and tree-planting campaigns. It has English-language signage, as well, and has hotels for tourists to stay in and visit old fortresses, dunes, oases, and grottos and even hire ultra-light aircraft to fly over the desert or visit Mongolia for a day. Jiuquan's main hotel sells Shenzhou memorabilia.

Travel from Jiuquan city to the launch site is via a smart blue-and-white train or on a dead straight road running parallel to it. At 1,000 m above sea level, Jiuquan offers unusually clear skies, almost clean-room conditions (interrupted only by occasional dust storms), and little rainfall. Skies like this make it possible to visually follow rockets far into the distance. Jiuquan is seasonal: very

hot in summer, with temperatures of up to +42.8°C, but these can drop to as low as -34°C in December. The countryside is made up of light, thin soil, small bushes, and camel thorn, with wildlife including goats and deer alongside domesticated camel and sheep. From time to time, soldiers clear dunes off the roads and railway lines. A rim of distant mountains can be seen to the north and west. Willow and poplar trees are planted over the cosmodrome for colour, shade, and to break up the wind.

The original cosmodrome covered 2,800 km² and was spread out across military, commercial, residential, and technical areas, connected by roads and rail, with a coal-powered power plant, a reservoir – whose water was used for drinking, growing vegetables, pig farms, and even orchards for vines and pears. Although American intelligence made its own sketches of the site, little was known of Jiuquan outside China until Swedish scientists and engineers travelled there for a satellite launch in 1992. The original 1970 launch site is now a preserved outdoor museum, but it is rusting and must be repainted every three years. It is possible to visit the underground control room and a ground-level museum with exhibits from the early days and ancient-looking cupboard-sized computers. Most of the soldiers are gone now.

Everything changed in 1992 with the crewed space programme, which required doubling the size of the cosmodrome southward (now 5,000 km²). The principal new feature was a vertical vehicle assembly building – in appearance not unlike that of Cape Canaveral – making it possible to do all the assembly vertically indoors and roll out a ready-to-go rocket to the pad, with fuelling the major task remaining before launch. This building was 92 m high, 27 m wide, and 28 m long,

Top:
The compact complex and its launch pads stick out against the flat plain of the desert, here looking out towards the mountain range to the south-east.
Source: Picture Alliance / dpa / Maxppp Xu Haihan

Right:
2012: Soldiers, personnel, and onlookers accompany the Long March 2F rocket to its launch pad. The rocket will later carry Chinese astronauts on their first manned space docking mission.
Source: Picture Alliance / Photoshot

with a 13-floor platform, cranes able to lift 17, 30, and 50 tonnes, and two high bays and two vertical processing halls, thereby able to prepare two launches at a time, with engineers having access from nine different levels. The building's door was 74 m tall, 8 m wide at the top, and 14 m wide at the bottom and weighed 350 tonnes, made of six 20-tonne sections. The building dominated the surrounding desert landscape and was visible 20 km away.

A horizontal transit building 78 m long by 24 m was built nearby. It was class 10,000, with air ducts to blow dust away. It was used to test out the launch vehicle. A non-hazardous operations building for spacecraft checking in clean room conditions and a hazardous operations building where fuels were loaded before launch were also built. Manned Shenzhou spacecraft are held in a 12-metre-tall scaffold before being lifted by a 15-tonne crane to the top of their rockets. The hall had motivational gold slogans printed on a red background. Connected by fibre optic cable 1,500 m from the pads was a 400 m² launch control centre equipped with a main control room with four rows of workstations and two smaller control facilities. There was a new accommodation area, called the western pavilion, where *hangtianyuan* ('astronauts') rest the night before their mission. The room where the first *hangtianyuan*, Yang Liwei, slept the night before his flight is preserved in perpetuity. This part of Jiuquan is well known to television viewers, who can watch *hangtianyuan* walk out, meet well-wishers, and board the bus to the pad.

To get to the pad 1,490 m away, the assembled rocket travelled on a crawler measuring 24 m in length, 21 m in width, and 8 m in height. It weighed 750 tonnes and was able to travel at 1.02 km/hr. Powered by eight electric motors, it took the crawler 40 minutes to travel from the vehicle processing building to the launch pad. There, the launcher and spacecraft were grappled by the umbilical tower – an 11-floor fixed steel structure 75 m tall, with floors for fuelling, electrical connexions, firefighting equipment, and an elevator. Underneath was an underground equipment room. There was a lightning conductor, flame trench, and a steel pipe down which hangtianyuan may slide in 60 seconds to a protected bunker in an emergency evacuation. The second pad was close by and branched just off to the left. It was first used only weeks after the first piloted mission, Shenzhou 5, when a CZ-2D put the FSW 3-1 recoverable satellite into orbit. At the new pad, the turnaround period was three days, meaning that a new rocket could be made ready for a mission within 72 hours of the previous launch.

Jiuquan progressed from being a secret, off-map facility in the 1960s to a tourist destination where visitors can now get tickets to crewed launches, although it is not clear exactly how. Up to 10,000 people crowd in, sharing four to a room in the old military accommodation. Because of its prominence in the piloted space programme and live broadcasting of the launches there, Jiuquan is the best known of the Chinese launch centres.

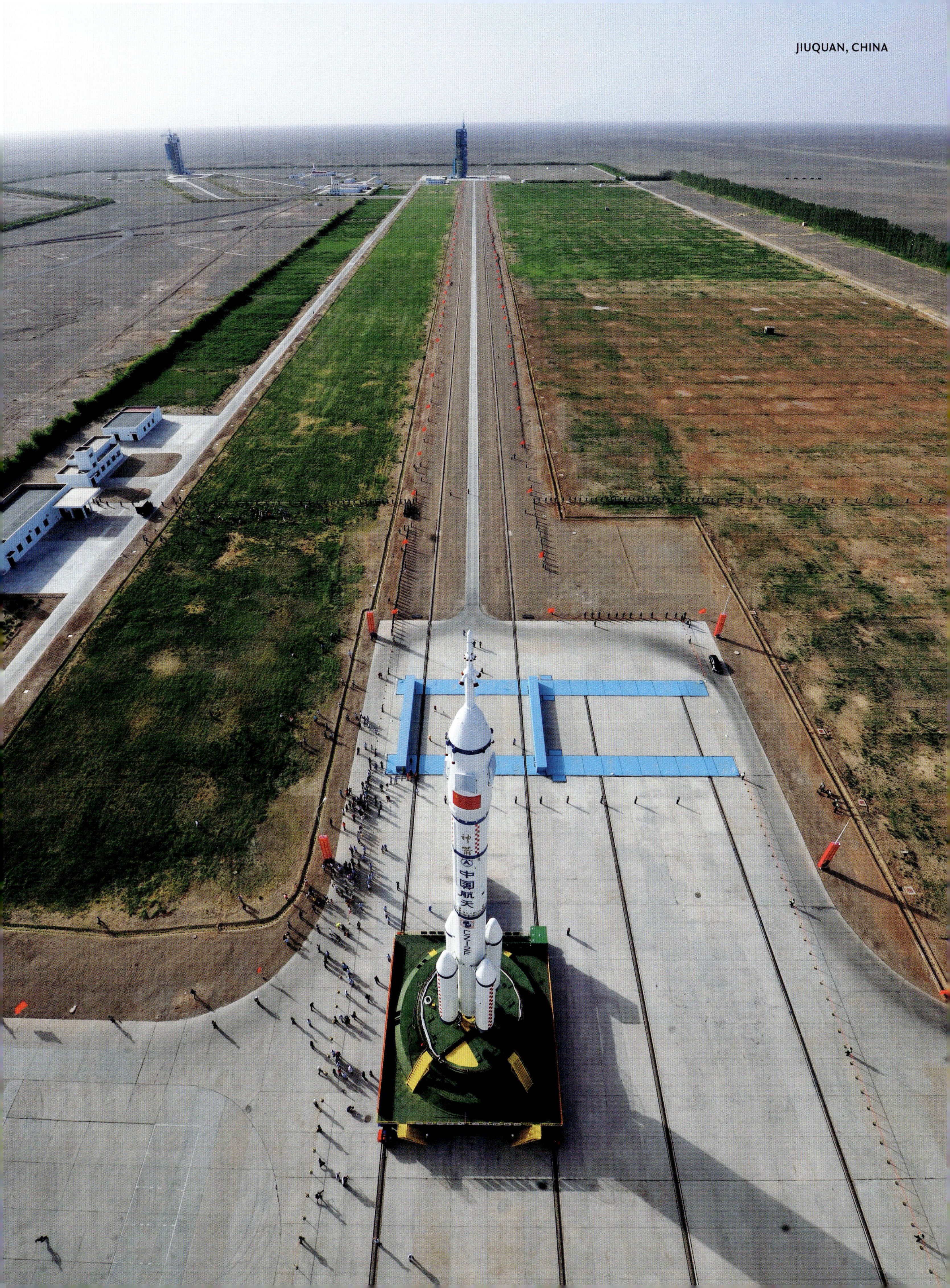
神舟
中国航天

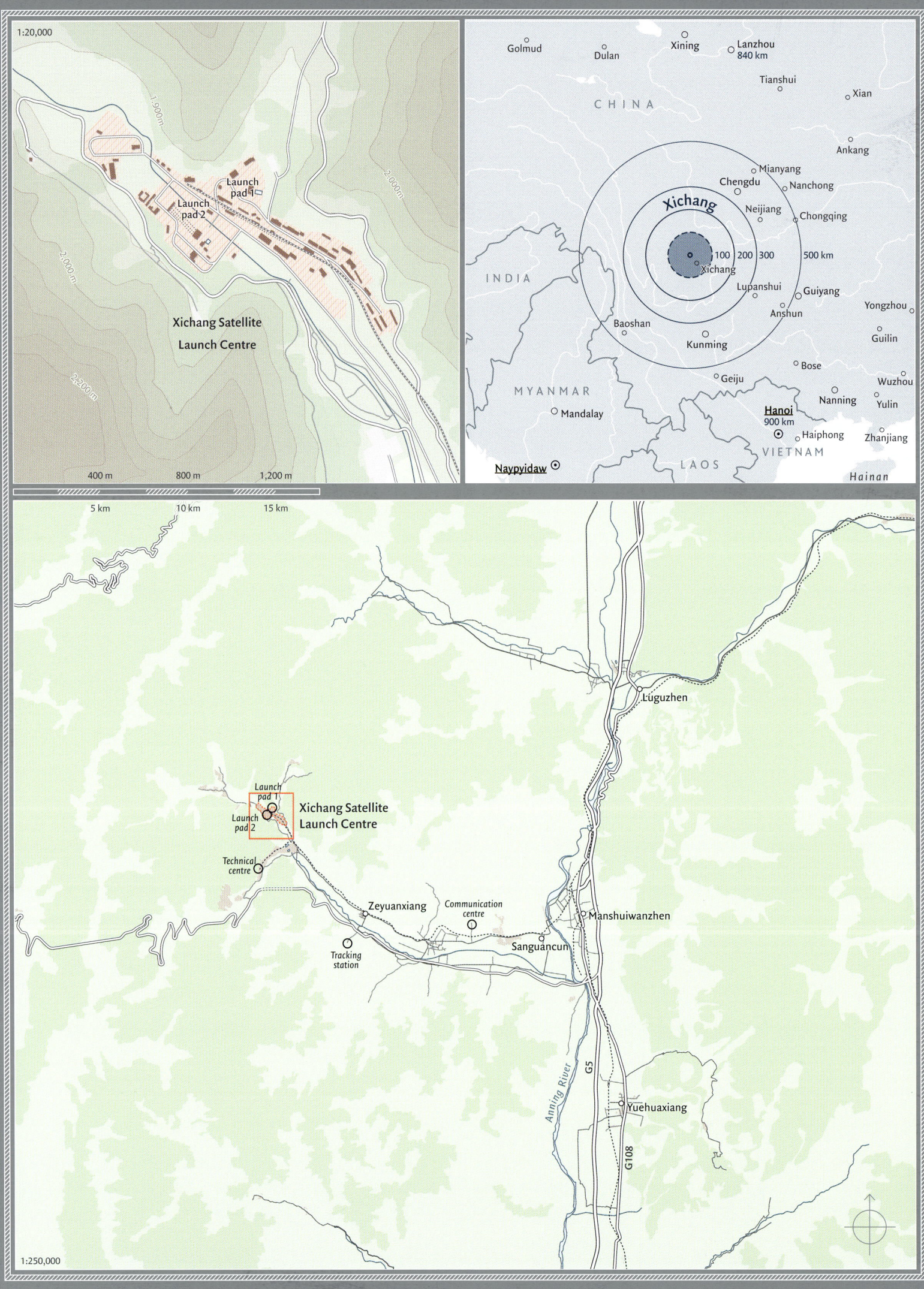

1:20,000

Launch
pad 1

Launch
pad 2

Xichang Satellite
Launch Centre

400 m   800 m   1,200 m

Golmud   Dulan   Xining   Lanzhou
840 km
Tianshui
CHINA   Xian

Ankang

Mianyang
Chengdu   Nanchong
Neijiang   Chongqing
INDIA   Xichang

100 200 300   500 km
Xichang
Lupanshui
Baoshan   Guiyang
Anshun   Yongzhou
Kunming
Guilin
MYANMAR
Geiju   Bose   Wuzhou
Mandalay   Nanning   Yulin
Hanoi
900 km
Haiphong   Zhanjiang
Naypyidaw   ⊙   LAOS   VIETNAM
Hainan

5 km   10 km   15 km

Lùguzhen

Launch
pad 1
Xichang Satellite
Launch
pad 2   Launch Centre

Technical
centre
Zeyuanxiang   Communication
centre   Manshuiwanzhen

Sanguancun
Tracking
station

G5

Anning River

Yuehuaxiang

G108

1:250,000

IN THE MISTY MOUNTAINS

Name Xichang Space Centre		**Coordinates** 28.3°N, 104°E
Location Xichang, China		**Time zone** GMT+8
Owner/Operator China National Space Administration		**Launches** 218 satellites
Elevation 1,542 m		**Completion** 1984

The misty mountains of Sichuan, southwest China, are the improbable location for its second launch site: Xichang. It was a relatively populated area, far inland, whose trajectory towards orbit brought it over some quite inhabited areas, compounded by poor communications. Although it enjoyed sunshine for most of the year (320 days) and a fairly constant, pleasant temperature (17°C), summers could see heavy rain and even flooding. For other reasons, though, the location made perfect sense. In 1969, China and the USSR came close to war, with border clashes between frontier troops, so a new location far from the existing Jiuquan (qv) launch site was needed. Secondly, when China decided to develop a programme of communications satellites to 24 hour equatorial, geostationary orbit, it needed a site much closer to the equator to get there. Zhou Enlai oversaw the process, reducing 80 prospective sites to 16 and then selecting Xichang himself.

The ground-breaking took place in December 1970 with a view to completion in two years (1972) and the site given the code name of '7201' (linked to 1972). It was first intended to be the launch site of the piloted Shuguang programme, but when that was cancelled in 1971, construction stalled. Work resumed in 1978 as the project for the communications satellite neared, the first pad being completed in 1982 and a launch rehearsal held there the following year. Black-and-white photographs show construction workers dressed in contemporary Zhongshan suits taking their lunch breaks on the hills as cranes rise in the background.

Xichang was the home of China's new rocket, the Long March CZ-3, based on but much more powerful than the Long March CZ-2, which already operated from Jiuquan. The CZ-3 had a hydrogen-fuelled upper stage and new engines that had the capacity to make the dogleg manoeuvre necessary to reach an orbit of 36,000 km.

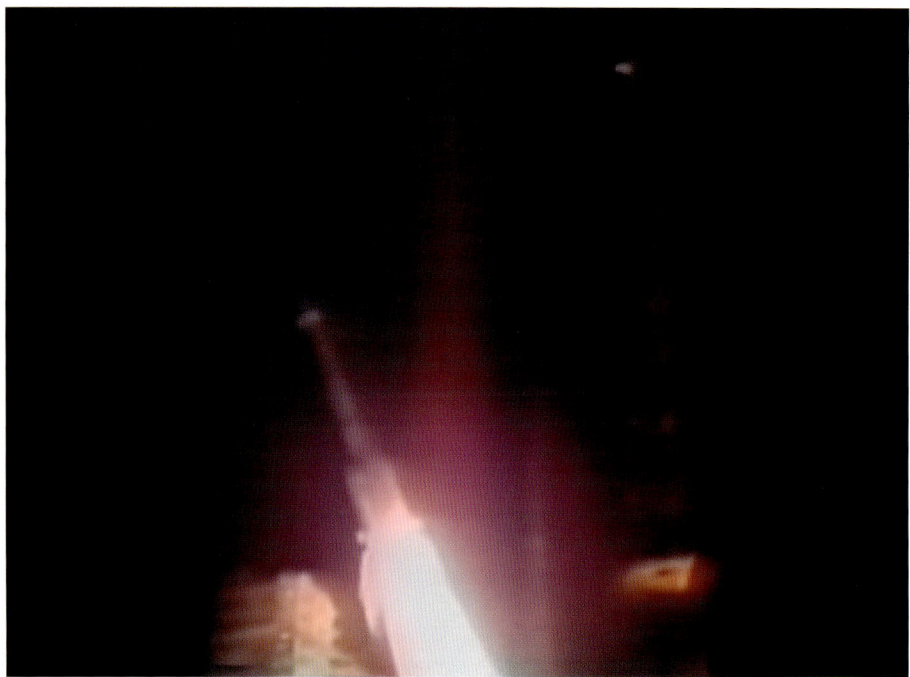

Top:
A collapsed building next to the hotel whose windows and doors had been blown out. American engineers on site stated that they witnessed large masses gathering around the now-devastated launch compound, while Chinese media reported only six dead.
Source: AirspaceMag / Bruce Campbell

Left:
During the launch of an American satellite, the Chinese CZ-3B steers off course turning the rocket into an unintentional cruise missile. The rocket impacted around 2.5 km down the valley.
Source: China Central Television

Although more powerful, hydrogen fuels must be kept extremely cold and required an additional level of sophisticated handling at the launch site. The first attempt failed in January 1984, but marked the inauguration of Xichang as a launch centre. A second attempt was made that April as summer rain and thunder threatened. This time, China successfully launched its first operational communications satellite, which became the basis of the modernisation of the country's communications. Xichang was one of the first locations in the Chinese space programme opened to foreign visitors. In the 1980s, China attempted to attract commercial business by launching Western communications satellites – a lucrative market, one which required lifting the American export blockade that had been in place since 1949. Once that was done, from 1990, Xichang began to welcome Australians and then Americans as the CZ-3 and a powerful derivative, the CZ-2E, launched their communications satellites. Visitors were struck by the location, over 1,800 m high in the mountains, which they reached by railway and roads that snaked along valley floors. Xichang was on the old southwestern Silk Road that started at Chengdu, 270 km away, and headed toward Burma and India. Visitors could see rice grow and buffaloes graze in fields adjacent to roads cluttered by bicycles and farm workers carrying chickens and vegetables to market. To the south there were lakes and to the north the big panda reserves. From time to time, the skies would fill with migrating birds. The Xichang launch site now has its own airport, connected to the launch centre by a railway, which in turn can reach Xichang city 65 km away on the Kunming-Chengdu high-speed line. The arrival of visitors did not turn out to have a happy ending. The first launch, attended by one hundred Australians and their live television crews staying in a purpose-built hotel, was successful in August 1991 and four more followed. In 1992 and 1995, though, in two events where the clients insisted on using an American kick motor, the satellite was lost, prompting much recrimination as to who was responsible. Worse was to follow on 14 February 1996 when the Long March CZ-3B failed two seconds after take-off on its inaugural mission. It tilted and exploded, all on live television. It was such a mighty explosion that little debris was found. The visitor hotel was demolished and there was an unknown number of casualties. Not only did foreign clients lose confidence in the programme, but the Americans accused the Chinese of spying on the Western satellites that they handled and re-imposed the 1949 blockade with ever-greater severity. It is still in force today.

That might have been the end for Xichang, but China got the CZ-3B airborne the following year and arranged to fly European satellites guaranteed not to have American components, thus beating the blockade. When the Americans put a stop to that by threatening the Europeans with sanctions as well, they offered the launch service to African, South American, and Asian countries for whom they provided a complete service of satellite, launcher, training, and finance. This and domestic programmes, especially navigation satellites (Beidou), made Xichang the busiest Chinese spaceport, overtaking Jiuquan. In 2007, Xichang became home to the first stages of the Moon programme with the launch of Chang e 1 to lunar orbit. Visitors paid ¥800 to watch the launch from 2,500 seats on observation platforms in the nearby hillsides.

Launches out of Xichang normally take a curving trajectory to the south-east, flying over southern Taiwan, north over the Philippines,

and towards the equator. The first launch pad was used by the CZ-3 and served the initial launches from 1984. Rockets and satellites were brought by rail from Beijing and then encased in a massive 900-tonne, 77-metre-tall gantry with 11 work levels and a crane to lift the parts in place. An air-conditioned clean room on the top floor protected satellites from dust and humidity. The countdown was carried out in a blockhouse close to the pad, but the overall operation and the subsequent flight were monitored from a launch control centre 6 km away. Built into the mountain, it comprised a large gymnasium-sized room with walls of consoles and a large, 4 m by 5.3 m visual display at the front. A cement flame trench was constructed to take away the rocket's flames on take-off.

After the CZ-3's last flight in 2000, the old tower was demolished and replaced by a new one weighing 1,800 tonnes and measuring 85.5 m in height. It was first used on the Beidou launch of 13 April 2007 and is now officially called Launch Complex 3 (LC-3). Chang e 1 took off from here in 2007 and it is normally used by the CZ-3A and CZ-3C. The command centre was renovated, the hotel rebuilt, and fresh staff accommodation provided – the first substantial renovation to the centre.

Xichang's second launch site (Launch Complex 2; LC-2) was built 350 m from the first and constructed in only 14 months – in what must be a world record – for the first CZ-2E launch in 1990. This second pad has a huge 4,580-tonne mobile service tower measuring 97 m in height, with 17 work levels and a fixed tower 74 m in height. Just 80 minutes before launch, the tower moves back to a distance of 130 m. After 1996, it was used mainly for the CZ-3B. Adjacent to the railway line for the second pad were buildings for

The dangers of launching over populated areas: villagers from the Gaopingsi Village in southwest China's Guizhou province surround the debris of CZ-3B, launched from Xichang in 2014.
Source: Picture Alliance / dpa / Wu Rubo

Next double page:
China sends two satellites of the Beidou Navigation Satellite System (BDS) into space from the Xichang Satellite Launch Centre in southwest China's Sichuan Province on 23 November 2019. Launched on a Long March-3B carrier rocket and the Yuanzheng-1 (Expedition-1) upper stage attached to the carrier rocket, the two satellites have entered their planned orbits. They are the 50th and 51st satellites of the BDS satellite family.
Source: Picture Alliance / Guo Wenbin

storing rockets, their various stages, and payloads, later assembled vertically on the pad by crane. The launch towers were protected by 100 m high lightning towers. Around the gantries were fuelling lines – one set to keep the liquid hydrogen third stage topped up, a second to provide helium that pressurises the fuel tanks, and another for storable fuels. Liquid hydrogen was topped up in the third stage until just three minutes before lift-off. There was an observation room capable of accommodating 500 people at a time. Laser theodolites, set in domes, tracked rockets as they ascended to orbit. A launch campaign in Xichang takes 40 days. The technical centre is 2.2 km from the pads. The rocket is first delivered by rail into a transit hall measuring 30.5 m by 14 m, before being brought into a much larger assembly room of 91.5 m by 27.5 m. Payloads are checked out in a clean room measuring 42 m by 18 m – called the non-hazardous operations building, where temperatures and humidity are kept within tight limits. The stages and payloads are then transferred to the hazardous operations and fuelling building where solid-rocket stages and satellites are installed. Final checks take place in a last checkout and preparation building. The site also has an X-ray facility to check any equipment for cracks. The rocket stages are then trolleyed to the pad, one by one, before being assembled vertically. Launch lists (e.g., for Beidou navigation satellites) indicate a level of interchangeability between LC-2 and LC-3. Xichang is now so well established as to have a museum, which includes a Shenzhou cabin – even though none were launched from here – and models of the Chang e spacecraft and other rockets.

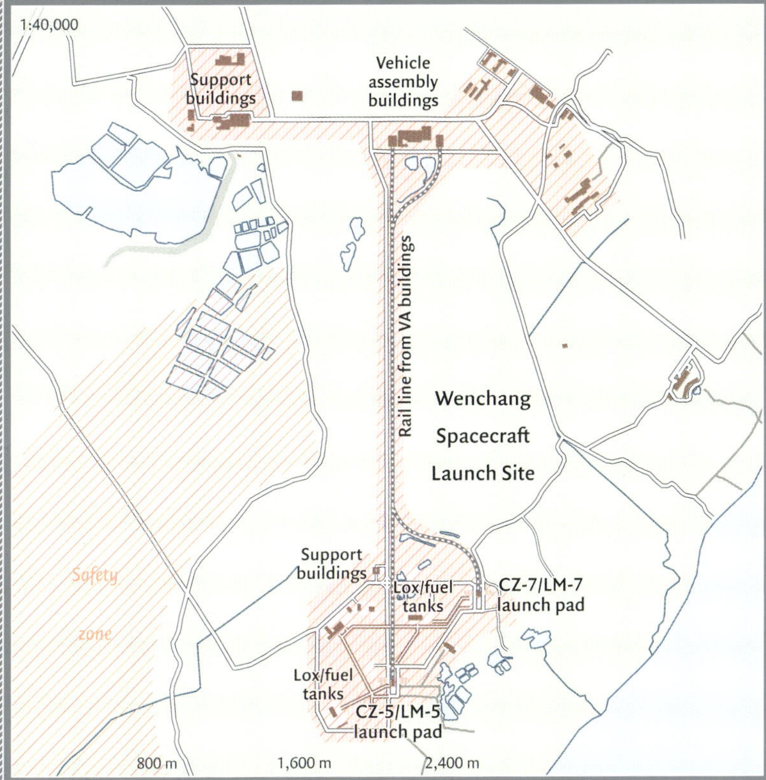
1:40,000
Support buildings
Vehicle assembly buildings
Rail line from VA buildings
Wenchang Spacecraft Launch Site
Support buildings
Lox/fuel tanks
CZ-7/LM-7 launch pad
Lox/fuel tanks
CZ-5/LM-5 launch pad
Safety zone
800 m
1,600 m
2,400 m

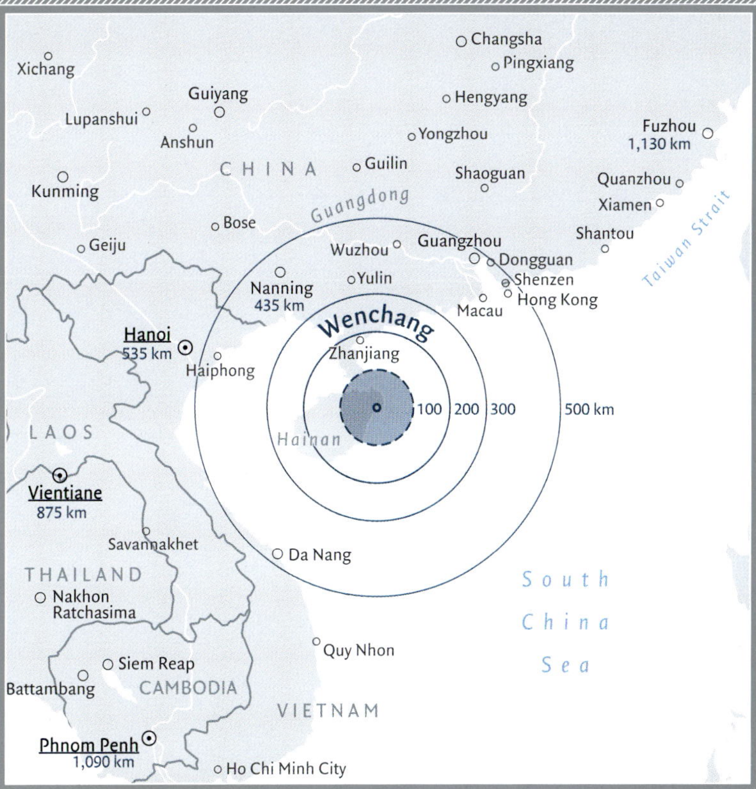
Changsha
Pingxiang
Xichang
Guiyang
Hengyang
Lupanshui
Anshun
Yongzhou
Fuzhou 1,130 km
CHINA
Guilin
Shaoguan
Quanzhou
Kunming
Xiamen
Guangdong
Bose
Shantou
Taiwan Strait
Geiju
Wuzhou
Guangzhou
Dongguan
Shenzen
Nanning 435 km
Yulin
Macau
Hong Kong
Wenchang
Hanoi 535 km
Zhanjiang
Haiphong
100
200
300
500 km
LAOS
Hainan
Vientiane 875 km
Savannakhet
Da Nang
THAILAND
Nakhon Ratchasima
South China Sea
Quy Nhon
Siem Reap
Battambang
CAMBODIA
VIETNAM
Phnom Penh 1,090 km
Ho Chi Minh City

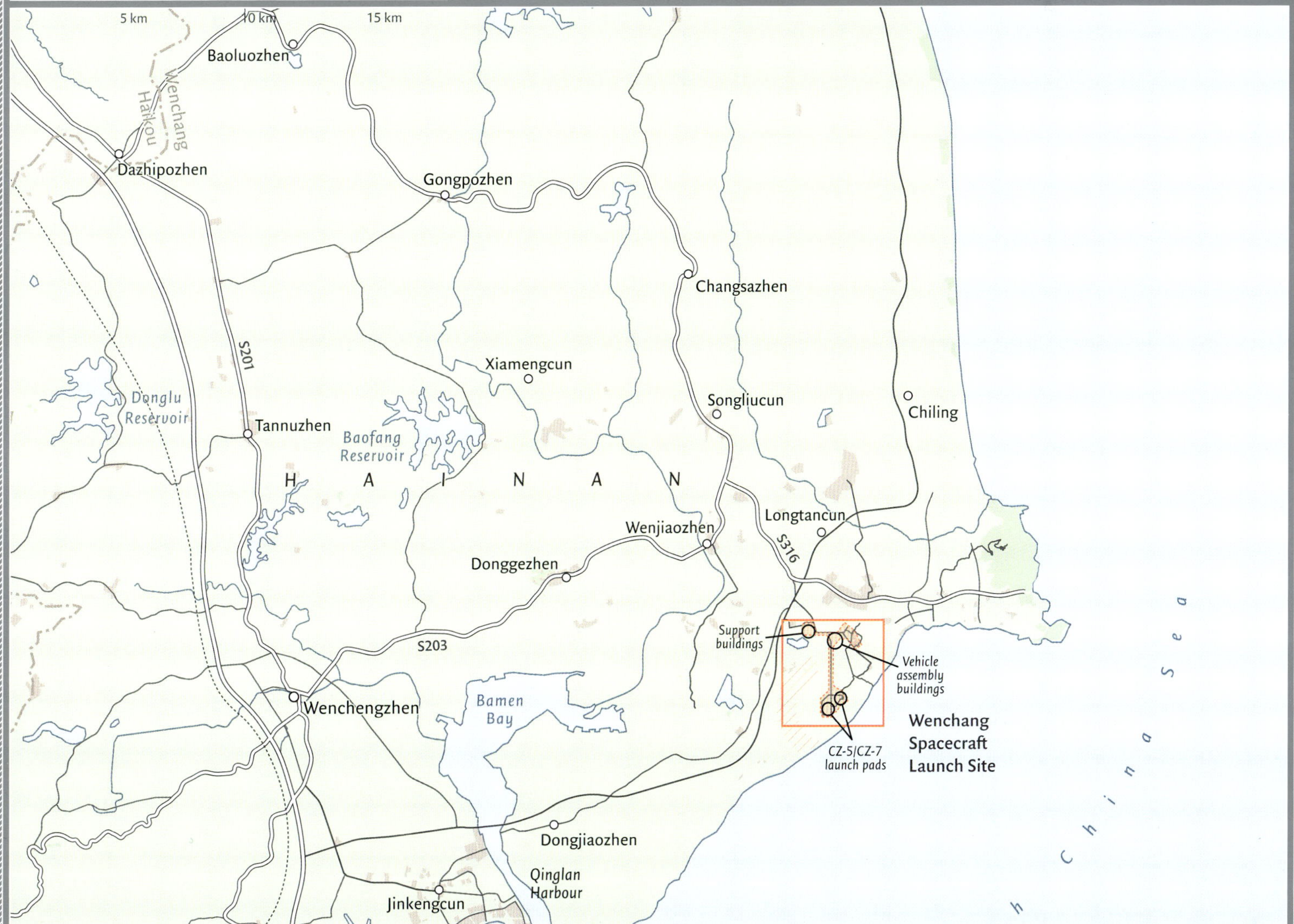
5 km
10 km
15 km
Baoluozhen
Wenchang Haikou
Dazhipozhen
Gongpozhen
Changsazhen
S201
Xiamengcun
Songliucun
Chiling
Donglu Reservoir
Tannuzhen
Baofang Reservoir
HAINAN
Longtancun
Wenjiaozhen
S316
Donggezhen
Support buildings
S203
Vehicle assembly buildings
Wenchengzhen
Bamen Bay
Wenchang Spacecraft Launch Site
CZ-5/CZ-7 launch pads
Dongjiaozhen
Qinglan Harbour
Jinkengcun
South China Sea
1:250,000

CHINA'S DESTINATION MOON LAUNCH SITE

Name
Wenchang Space Launch Site

Location
Hainan, China

Owner/Operator
China Aerospace Science and
Technology Corporation

Elevation
13 m

Coordinates
19.3°N, 111.5°E

Time zone
GMT+8

Launches
45 satellites

Completion
2016

Wenchang is China's newest launch site, completed at around the same time as Russia's modern cosmodrome, Vostochny (qv). China already had launch sites in Jiuquan (qv), its first, Xichang (qv) in Sichuan for equatorial orbit, and Taiyuan (qv) for applications satellites. So why build another? The reason was simple enough: railways. China was building much bigger rockets – for its space station (the CZ-5, CZ-7) and for its manned lunar landing missions (CZ-9). Their lower stages were simply too large for the railway system, which could take stages up to 3.35 m in diameter, but the CZ-5 was 5 m in diameter and the CZ-9 10 m. They required sea transport in the form of two special cargo ships, the Yuanwang 21 and Yuanwang 22, bringing the components all the way from Tianjin in north-east China, which is the port of Beijing, connected by canal. In effect, the architecture of Wenchang is intimately associated with that of Tianjin, where the

CZ-5, 7, and 9 and their variants are made. Both facilities were built and completed at around the same time – the mid-2010s. The rocket stages are loaded onto the Yuanwang 21 and Yuanwang 22 in Tianjin Port to make the seven-day sea journey to Qinlang, from where they are brought to Wenchang. Construction of space facilities in Binhai, Tianjin, began in 2009. Its size matched that of the ambition of the rockets made there, being 313 ha and costing €900m. The manufacturing floor covers a vast area of 220,000 m², where up to 12 rockets can be built at a time. The payload fabrication factory has a floor of 100,000 m² and is able to handle eight spacecraft at a time, with an Assembly, Integration, and Testing (AIT) building measuring 99 m in height and a steel door measuring 81 m in height.

Wenchang was selected as a launch site in 2000, the precise location being Longlou on the Tonggu Jiao peninsula, jutting out from

the north-eastern shore of Hainan Island. The island itself is not well known, although it is China's largest. It was considered a backward province until it became an experiment in rapid, unregulated capitalist modernisation from the 1980s. It is separated from the mainland by a narrow strait and to the west is Vietnam. It was in the Western news briefly in 2001 when an American EP-3 spy plane was forced down there and its crew interned. Hainan had a previous connection to the space programme because a sounding rocket site was built earlier near Haikou on the western side of the island. Five Weaver Girl sounding rockets were fired from here between 1988 and 1991, with missions resuming in 2011. Leaving aside the question of outsize components, Wenchang was closer to the equator, offering a 7.4 per cent payload advantage on Xichang. It launched straight out over the sea at a time when stages from the other sites were falling on populated areas and people's homes, attracting adverse media publicity. Wenchang was a city of 520,000 – small by Chinese standards – and 6,000 people had to be relocated for the launch site. In China itself, the island was becoming better known for tourism (it boasted its own airline, Hainan Airlines) and as a retirement home for the growing number of affluent elderly. With its sandy beaches, coconut palms, mangrove forests, lagoons, Confucian temples, and year-round warmth, it was well-suited for both.

Wenchang's construction began simultaneously with Tianjin. The soil turning ceremony was on 14 September 2009, with prominent media publicity with images of cranes, cement mixers, and hard hats. The launch area site measured 20 km² and was only 3 km from the coast. The new site, initial cost €500m, was strongly promoted and part-financed by the Hainan regional government, not unusual in modern China, which hoped to attract foreign investment (e.g., Japan), high-tech industries, and tourism. The site included two downrange island tracking stations: Tongguling, 5 km to the east, and the Xisha Islands (also called the Paracel Islands), far to the southeast. To receive the rockets from Tianjin, Wenchang Port was extended with a new pier and hydraulic lifting facilities. The parts were lifted out for a slow, 10-hour outsize convoy load road journey. Wenchang has two launch pads: one for the CZ-5, called pad LC101, and the other for the CZ-7, called LC-201. There is a double-door vehicle assembly building measuring 101 m in height – 8 m higher than Jiuquan. It has a 2,800-metre reddish sand trackway down to the pad, with the rocket carried on a 2,000-tonne mobile launch platform – 27.5 m tall for the CZ-5 and 23 m tall for the CZ-7. They were supplemented by 58-metre-tall umbilical towers providing facilities for fuelling, gas, air conditioning, and electrical power; and four lightning towers. Each pad had a large flame diversion trench.

Wenchang's inaugural launch was the CZ-7 on 25 June 2016, with the second CZ-7 sending the Tianzhou 1 freighter up to the Tiangong 2 space station on 20 April 2017. Unsurprisingly, it was the much bigger CZ-5 that attracted the most interest. Its first launch took place on 3 November 2016, with enthusiastic crowds cheering as it headed out over the sea, its bright yellow flame reflected on the sea offshore. Big crowds were a feature of CZ-7 and CZ-5 launches, which for some reason seemed to happen mainly at night. The large grass viewing area has a big screen, musical entertainment before the launch, food, and pancake and drink stands. Lightly dressed for the tropical conditions, almost everyone in the crowd holds up camera phones to record the dazzling rockets ascending skyward as the sound washes over them, almost blowing them over.

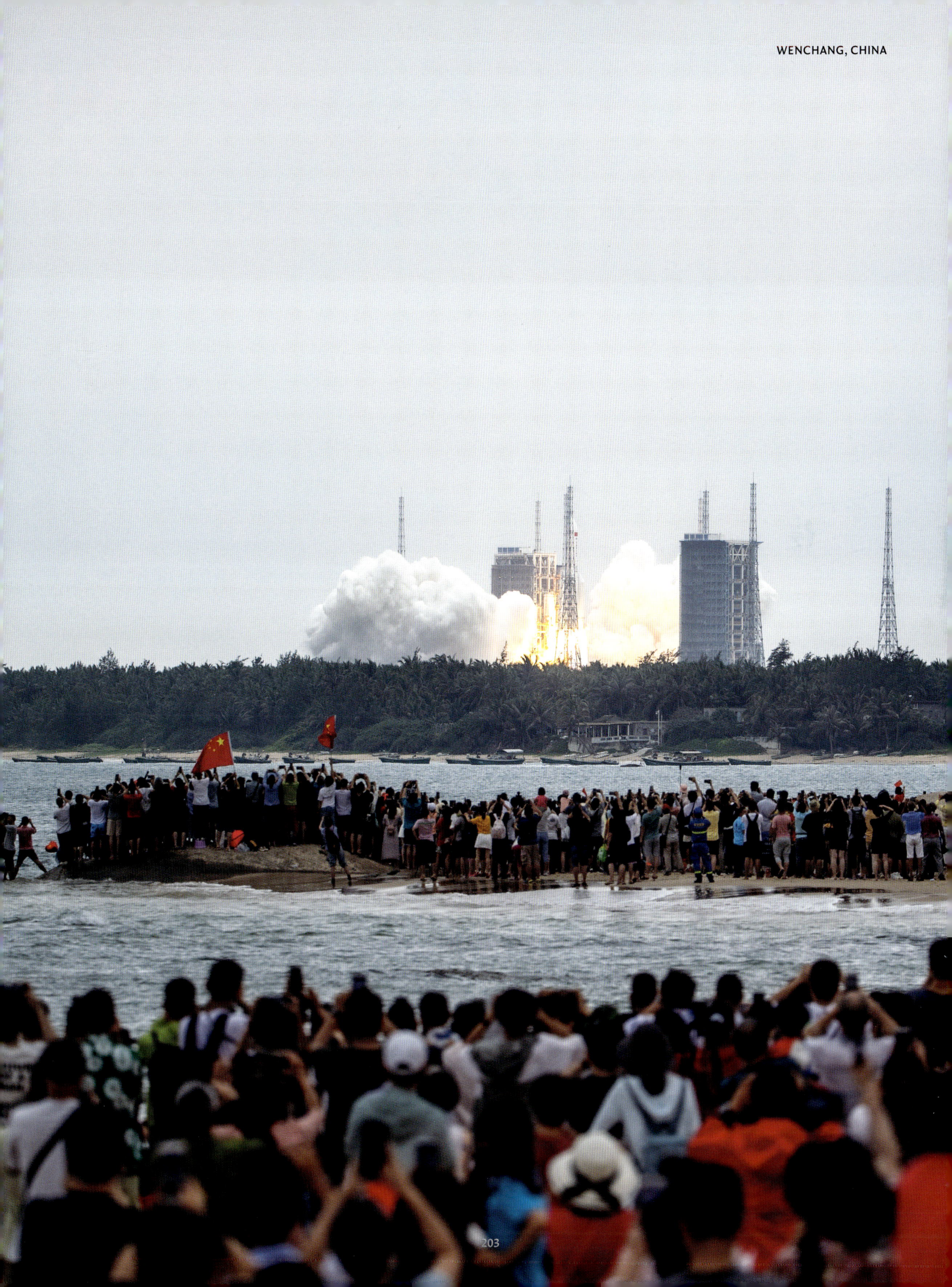

A CZ-7 carrier rocket moves vertically to the launch
tower crawling along the long avenue, accompanied
by a swarm of personnel.
Source: Picture Alliance / Xinhua / Guo Cheng

Left:
A Long March-8 Y1 rocket at the Wenchang Spacecraft
Launch Site. The rocket was vertically transported
to the launching area before being prepared for launch in
December 2020.
Source: Picture Alliance / Xinhua / CNSA

Next double page:
Construction site of one of the buildings of China's Wenchang
International Aerospace City project (2021).
Source: Picture Alliance / Xinhua / Guo Cheng

The second version of the CZ-5, the CZ-5B, was launched from
here on 5 May 2020, watched by large, excited crowds lining the
choppy waters of the nearby shoreline. This is a shorter, stubbier
version, able to lift over 20 tonnes into Earth orbit, on this occa-
sion orbiting a prototype 20-tonne spaceship that will later bring
hangtianyuan to the Moon. The spaceship, the Xinyidai Zairen
Feichuan Shiyan Chuan, reached an altitude of 8,000 km before
being driven into the Earth's atmosphere at high speed to practice
a return from the Moon, landing intact on the grasslands of inner
Mongolia. This is the version that will lift China's space station's
modules: the Tianhe, Mengtian, and Wentian. Wenchang was and
is the most open of the Chinese launch sites, with a visitor centre
opened before the inaugural launch. Its outdoor exhibits comprise
a full-scale model of Tiangong docked with Shenzhou. Inside, the
Shenzhou 10 cabin and a spacesuit are on show, along with models
of the Yutu lunar rover, the Tianwen 1 Mars 2020 lander and rover,
and the space station. Visitors pay about €18 and can drive around
on battery-powered carts to see the launch and control centre, vehicle assembly building, and launch tower. They may stay in international hotels close by (Hilton) or with local families, boost-
ing local soft tourism and the incomes of fishing villages, which
have been tidied up. The local government plans both tourist facilities (e.g., space-themed amusement parks, hotels and souvenirs, a
4D cinema and discovery hall) and high-tech industries connected
to the space programme. It is hard to see this being a difficult sell.
It is primarily geared to Chinese visitors, but Western visitors fly
to Haikou, take the high-speed train to Wenchang railway station
(20 minutes), and navigate the facilities from there.

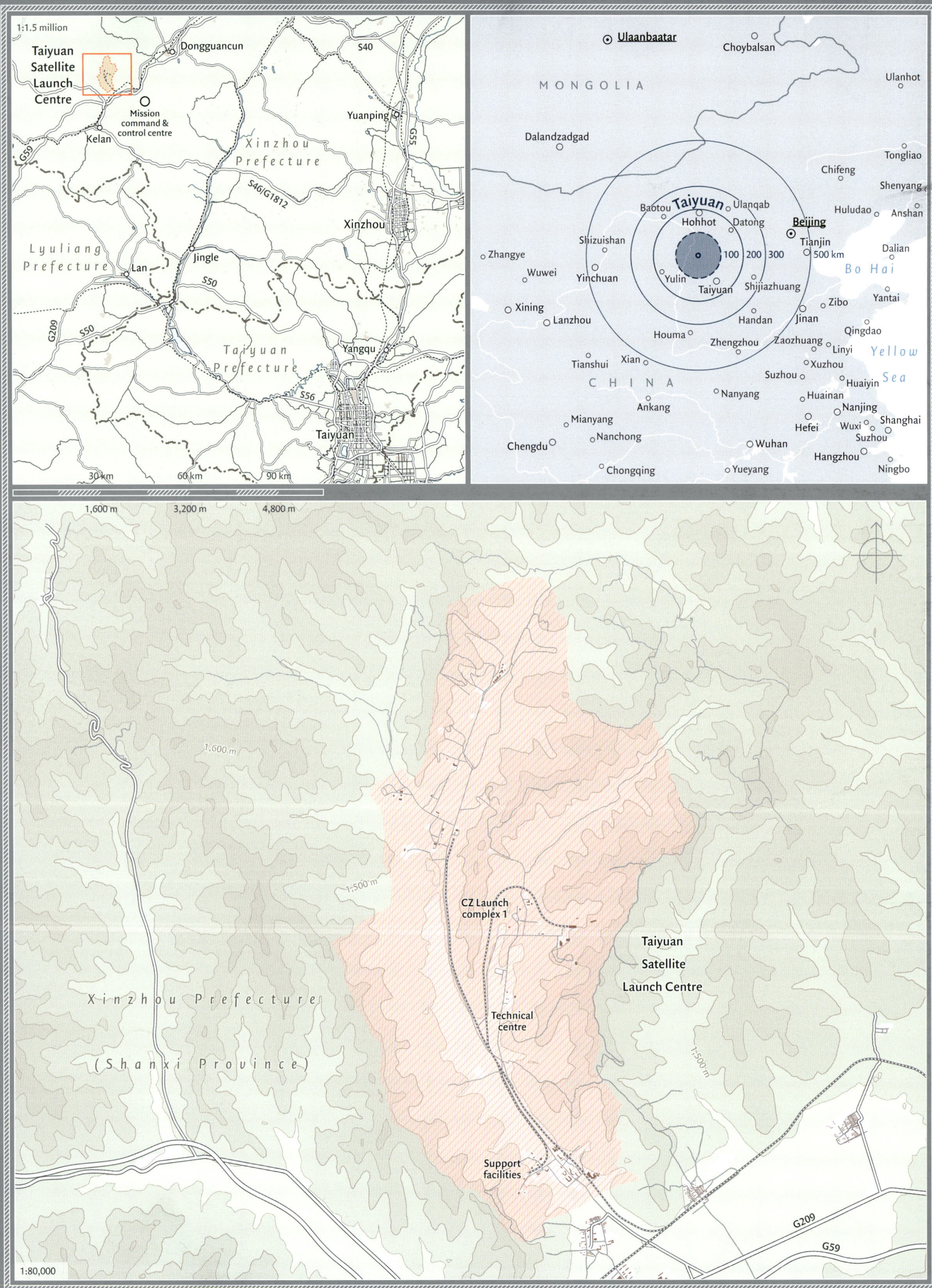

Taiyuan
Satellite
Launch
Centre

Dongguancun

S40

Mission
command &
control centre

Kelan

Xinzhou
Prefecture

G59

S46/G1812

Yuanping

G55

Xinzhou

Lyuliang
Prefecture

Lan

Jingle

S50

Taiyuan
Prefecture

Yangqu

S50

G209

S56

Taiyuan

30 km   60 km   90 km

Ulaanbaatar   Choybalsan

MONGOLIA   Ulanhot

Dalandzadgad   Tongliao

Chifeng   Shenyang

Taiyuan   Ulanqab   Huludao   Anshan

Baotou   Datong

Hohhot   Beijing   Dalian

100 200 300   Tianjin

Zhangye   Shizuishan   Yulin   500 km   Bo Hai

Wuwei   Yinchuan   Taiyuan   Shijiazhuang   Yantai

Xining   Zibo   Qingdao

Lanzhou   Houma   Handan   Jinan   Yellow

                        Zhengzhou   Zaozhuang   Linyi   Sea

Tianshui   Xian   Suzhou   Xuzhou   Huaiyin

CHINA   Nanyang   Huainan   Nanjing

Mianyang   Hefei   Wuxi   Shanghai

Ankang   Nanchong   Suzhou

Chengdu   Wuhan   Hangzhou   Ningbo

Chongqing   Yueyang

1,600 m   3,200 m   4,800 m

1,600 m

1,500 m

CZ Launch
complex 1

Taiyuan
Satellite
Launch Centre

Xinzhou Prefecture

Technical
centre

(Shanxi Province)

1,500 m

Support
facilities

G209

G59

THE COAL TOWN SITE

Name		**Coordinates**
Taiyuan Satellite Launch Centre		39.5°N, 112.6°E
Location		**Time zone**
Taiyuan, China		GMT+8
Owner/Operator		**Launches**
China National Space Administration		231 satellites
Elevation		**Completion**
800 m		1988

Taiyuan is unmistakeable, especially in winter, when rockets ascend from its low, snow-sprinkled, drumlin-like hills billowing clouds of brown smoke. It is probably the least known of China's launch sites, located in the north-eastern, colder part of the country in Kelan County southwest of Beijing. It is 1,500 m above sea level near the towns of Taiyuan and Wuzhai in an area known for coal production. Winter temperatures can be as low as -30°C. Historically, it was a military base (Site 21), with a single pad used to test the Dong Feng missile from 1968 onwards. Taiyuan was brought into the space programme both because other launch sites had become ever busier and because it was suitable for launches into polar orbit. A 76.9-metre-tall, 11-floor launch tower was installed, not unlike the one in Jiuquan for the CZ-2. Taiyuan began as the home of the new Long March CZ-4 launcher, used to send the Fengyun 1 and 3 series of weather satellites into polar orbit from 6 September 1988 (Fengyun 1-1) onwards. The CZ-4 pad has a fixed service tower with rotatable service platforms and crane. Over time, Taiyuan was used for military satellites (e.g., the Yaogan series and Shijian 6 series) and others for Earth observation (e.g., Ziyuan, Huanjing, Haiyang, Gaofen, Gaojing) and technology testing (e.g., Shijian 9). The launch facilities were sufficiently compatible for the slightly smaller CZ-2 to be launched from there. As Taiyuan was used more, a second pad was installed in 2008. Although visitors have been to Taiyuan to accompany their payloads (e.g., Tan Ce 2), there is a limited photographic record, and our main knowledge comes from Brazilian space officials accompanying their China–Brazil Earth Resources Satellites (CBERS) launches.

Behind the barbed wire, the bright blue launch
tower hosts China's Long March 4C rocket.
Source: Picture Alliance / dpa / HPIC

On 20 September 2015, Taiyuan became home to the new, small
Long March CZ-6. This required very few adaptations, for the
CZ-6 could be towed on a powerful military trailer along wind-
ing roads and up to a flat hilltop pad, raised to the vertical, and
fired. The CZ-6 marked the start of the replacement of China's old
launcher fleet (CZ-2, 3, and 4) with the new (CZ-5, 6, 7, 11, and 9).
Site facilities now include a railway station, two pads (CZ-4, also
used for CZ-2; CZ-6), technical centre, launch control, tracking
systems, and a power station.

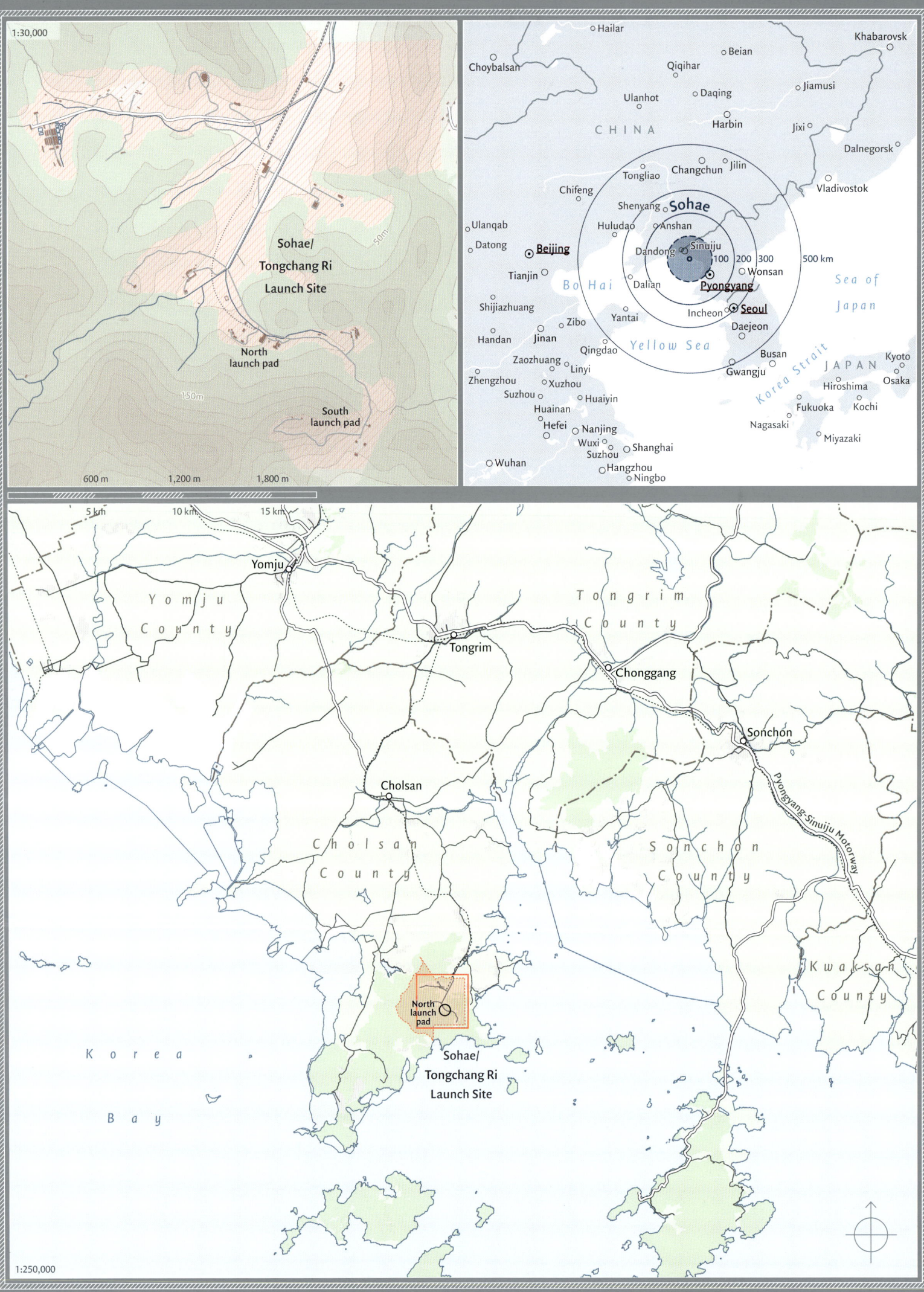

1:30,000

Sohae/
Tongchang Ri
Launch Site

North
launch pad

South
launch pad

600 m   1,200 m   1,800 m

Hailar
Choybalsan
Khabarovsk
Beian
Qiqihar
Ulanhot   Daqing   Jiamusi
CHINA   Harbin   Jixi
Dalnegorsk
Chifeng   Tongliao   Changchun   Jilin
Shenyang   Sohae   Vladivostok
Ulanqab   Huludao   Anshan
Datong   Beijing   Dandong   Sinuiju
100   200   300   500 km
Tianjin   Dalian   Wonsan
Bo Hai   Pyongyang   Sea of
Shijiazhuang   Yantai   Incheon   Japan
Handan   Jinan   Yellow Sea   Seoul
Zibo   Daejeon
Zaozhuang   Qingdao   Busan
Zhengzhou   Linyi   Gwangju   Korea Strait   Kyoto
Suzhou   Xuzhou   JAPAN   Osaka
Huainan   Huaiyin   Hiroshima
Hefei   Nanjing   Fukuoka   Kochi
Wuxi   Nagasaki
Wuhan   Suzhou   Shanghai   Miyazaki
Hangzhou
Ningbo

5 km   10 km   15 km

Yomju

Yomju
County

Tongrim
County

Tongrim

Chonggang

Sonchon

Cholsan

Cholsan
County

Sonchon
County

Pyongyang-Sinuiju Motorway

Kwaksan
County

North
launch
pad

Korea

Sohae/
Tongchang Ri
Launch Site

Bay

1:250,000

DPRK'S SITE ECHOES POLITICS OF KOREAN PENINSULAR

Name
Sohae Satellite
Launching Station

Location
North Pyongan, Democratic
People's Republic of Korea

Owner / Operator
North Korea space
programme

Elevation
60 m

Coordinates
39°N, 124°E

Time zone
GMT+9

Launches
2 satellites

Completion
2012

The two Koreas are among the more recent space powers, achieving their first launches not far apart: 2012 for the north (Democratic People's Republic of Korea; DPRK) and 2013 for the south (Republic of Korea). Because the north's space programme has been linked to missile development, its launch sites have attracted a high level of intelligence attention. Indeed, they have been described as the world's most photographed secret launch sites. The DPRK initiated a missile programme in 1966, obtaining assistance from the Soviet Union and China in the development of medium-range Scud missiles known as Nodong. From 1993, a three-stage version, Taepo Dong 1 (also called Unha 1), was developed to launch small satellites. The rocket weighed 20 tonnes and measured 25.5 m in length and was able to lift 25 kg to low Earth orbit. The DPRK first claimed to have orbited a satellite as far back as August 1998, with the launch site being identified in the north-east tip of the country: Musudan-ri, Hwadae County, North Hamgyong province, also called Tonghae (40°13'N, 149°07'E). Another attempt was made in April 2009. Musudan-ri was a rudimentary site in an isolated rural, hilly region, with personnel driving in on unpaved roads to fire missiles and rockets. From 2012, launches were moved to Sohae at the village of Pongdong-ri, Cholsan County, North Pyongan province in the northwest of the country in wooded hills near the Chinese border. Construction began in the early 2000s and was completed in 2009 when its existence was announced. With the advantage of being closer to railways, this was a much larger (6 km²), permanent facility with a launch stand, 10-level service tower, static motor test stand, launch control, fuel bunkers, and support buildings.

The first satellite launch from here used the Unha 3 rocket, which measured 30 m in height and 2.4 m in diameter and weighed 91 tonnes. It failed in April 2012, but success was finally achieved on 12 December 2012, putting the 170 kg observation satellite Kwangmyŏngsŏng 3 ('bright star') into orbit. Kwangmyŏngsŏng 4 followed in February 2016. Following the US-DPRK 2018 demilitarisation agreement, Sohae was dismantled, but when the next summit failed the following year, satellites observed reconstruction activity and it was operational again by 2020.

Top:
DPRK soldiers flank the Unha 3 rocket ready to be launched. Although the satellite onboard is launched for peaceful purposes, the rocket itself can also be used as a ballistic missile, as is the case with other countries.

Bottom:
Under the watchful eyes of the dictator, DPRK engineers test a rocket engine on a test stand at Sohae.

Right:
The line between satellites and intercontinental ballistic missiles remains blurry in the DPRK. The government attempts to keep them separated, geographically launching its ICBMs from undisclosed locations in the mountainous terrain of the country.

Source: Korean Central News Agency

Reporters trekking back from the launch pad.

Source: Korean Central News Agency

2012: The DPRK's Unha-3 rocket lifts off from the Sohae launch pad in Tongchang-ri. When the missiles are not standing vertically on the test stand or on the launch pad, they can be seen at Pyongyang's military parades.
Source: Korean Central News Agency

Top:
A satellite image provided by DigitalGlobe and captured on 2 March 2019 shows the launch tower at the Sohae Satellite Launch Facility in Tongchang-ri, Democratic People's Republic of Korea. The DPRK has been restoring facilities at the long-range rocket launch site it dismantled in 2018 as part of disarmament steps, according to foreign experts and a South Korean lawmaker who was briefed by Seoul's spy service. The finding followed a high-stakes nuclear summit in February 2018 between Supreme Leader of the DPRK Kim Jong Un and the then US President Donald Trump that ended without any agreement.
Source: Picture Alliance / AP Photo / Digitalglobe

Right:
This image provided by DigitalGlobe shows what is believed to be the exhaust trail and vehicle launched from Musudan-ri, DPRK, on 5 April 2009. The DPRK media claimed on 7 April 2009 that a rocket launch seen overseas largely as a technical failure will propel the country to greatness, while moves at the United Nations to punish Pyongyang remained mired in a lack of consensus. The DPRK launched what it claims was a satellite that successfully entered orbit around the Earth, defying international warnings that the move would violate UN resolutions and invite further censure.
Source: Picture Alliance / dpa / epa Digitalglobe

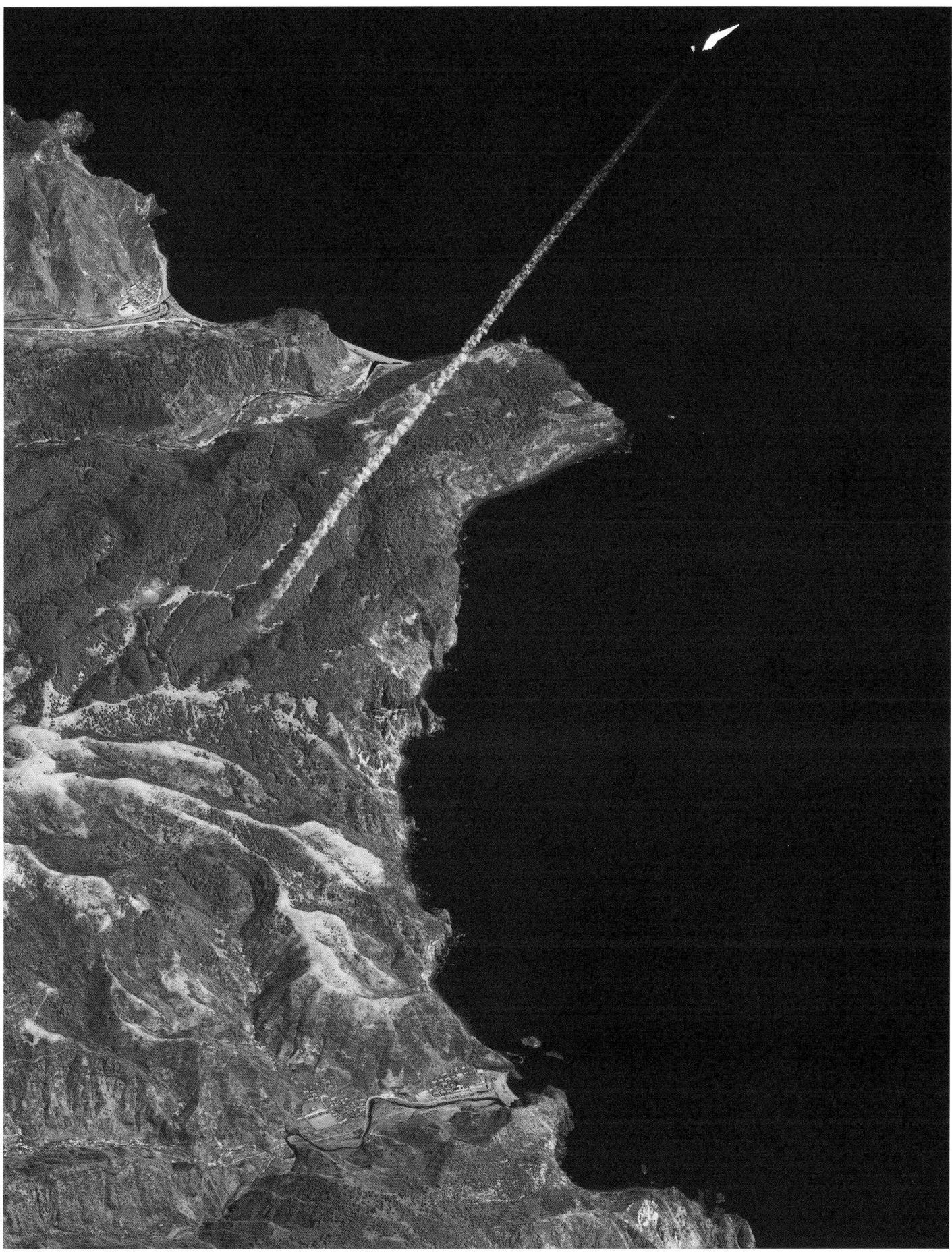

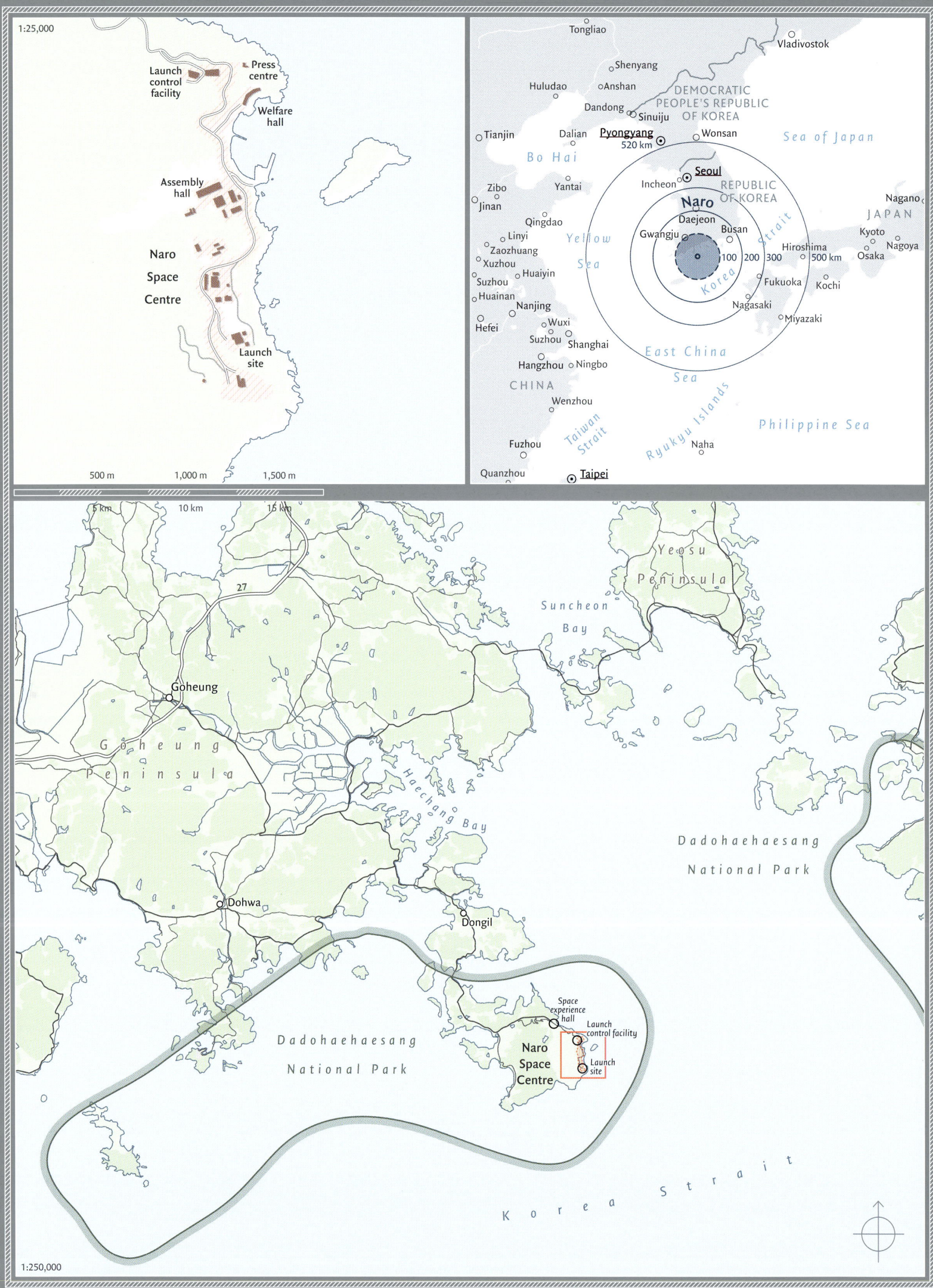

1:25,000

Launch
control
facility

Press
centre

Welfare
hall

Assembly
hall

Naro
Space
Centre

Launch
site

500 m   1,000 m   1,500 m

Tongliao

Vladivostok

Huludao

Shenyang

Anshan

DEMOCRATIC
PEOPLE'S REPUBLIC
OF KOREA

Dandong

Sinuiju

Wonsan

Tianjin

Dalian

Pyongyang

520 km

Sea of Japan

Bo Hai

Seoul

Zibo

Yantai

Incheon

REPUBLIC
OF KOREA

Nagano

Jinan

Yellow

Naro

JAPAN

Qingdao

Daejeon

Busan

Kyoto

Nagoya

Linyi

Gwangju

Hiroshima

Osaka

Zaozhuang

Sea

100 200 300   500 km

Xuzhou

Huaiyin

Fukuoka

Kochi

Suzhou

Huainan

Nanjing

Nagasaki

Hefei

Wuxi

Miyazaki

Suzhou

Shanghai

East China

Hangzhou

Ningbo

Sea

Wenzhou

Philippine Sea

CHINA

Fuzhou

Taiwan
Strait

Ryukyu Islands

Naha

Quanzhou

Taipei

5 km   10 km   15 km

Yeosu
Peninsula

27

Suncheon
Bay

Goheung

Dadohaehaesang

National Park

G o h e u n g

P e n i n s u l a

Haechang Bay

Dohwa

Dongil

Space
experience
hall

Dadohaehaesang

Launch
control
facility

National Park

Naro
Space
Centre

Launch
site

K o r e a   S t r a i t

1:250,000

Naro, Republic of Korea

KOREA'S ISLAND LAUNCH SITE

Name
Naro Space Centre

Location
Jeollanam-do,
Republic of Korea

Owner / Operator
Korea Aerospace
Research Institute

Elevation
10 m

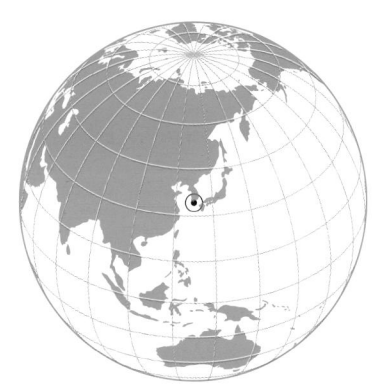

Coordinates
34.4°N, 127°E

Time zone
GMT+9

Launches
1 satellite

Completion
2013

The two partitioned countries are among the more recent space powers, achieving their first launches not far apart: 2012 for the DPRK and 2013 for the Republic of Korea. The Republic's initial interest and expertise was in small satellites for high-resolution Earth observation, before it went on to develop its own launcher and launch site. The launch site is Naro on the wooded, hilly island of Oenaro (45 km²; population 4,000) in the southwest of the country, 485 km south of Seoul. Development of a launch vehicle, the Korean Satellite Launch Vehicle I (KSLV I; also called Naro I), and a launch centre – the Space Rocket Complex (SRC) (also called Naro) – was a Korean–Russian project dating back to 2004. The scenic seaside launch site was built by the Russian Khrunichev rocket company, Hyundai Heavy Industries, and KBTM, a leading construction company for the Russian space programme. The 5.3 km² site included

a launch pad, erector, an integration and assembly building, storage and supply facilities, accommodation, offices, a tracking station, and a mission control centre 2 km away. There is a big educational visitor science centre with a museum. Naro rockets ascend over the Pacific towards Okinawa on a pathway careful to avoid the main Japanese archipelago. For the KSLV I, Russia supplied the first stage based on its new Angara rocket and the engine, the RD-151 – a powered-down version of the Angara's RD-191 engine. For the first launch, the first stage was delivered by an Antonov 124 cargo plane to the city of Busan, put on a ship in its large harbour, and transported to the island. The two-stage launcher weighed 140 tonnes, measured 33 m in height and 2.9 m in diameter, and was designed to orbit a 100 kg satellite. The second stage was a domestic solid-fuel rocket. The first launch took place on 25 August 2009, but the

Left:
The complexities of the rugged terrain become more than apparent when bringing the rocket from the assembly complex to its launch pad via the windy road.
Source: Picture Alliance / dpa / Str

Top:
Smoke produced by a 75-tonne liquid-fuel rocket engine firing test conducted at the Naro Space Centre drifts out to sea. The main launch pad that represents the Republic of Korea's gateway to space is visible in the background.
Source: Korea Aerospace Research Institute

satellite remained bolted to the Korean upper stage and crashed. A second launch in 2010 exploded. It was third time lucky on 30 January 2013 when Science and Technology Satellite 2C was put into orbit. The centre was subsequently expanded for the indigenous KSLV II. A second pad and service tower were added and new engine test facilities were cut into a high promontory. The KSLV II was successfully launched on a suborbital mission on 28 November 2018 and finally reached orbit in June 2022.

October 2012: The launch vehicle leaves the final assembly
area to begin the journey to the launch pad.
Source: Picture Alliance / dpa / Yonhap

Right:
The Nuri rocket, the first domestically
produced space rocket, lifts off from
a launch pad at the Naro Space Centre
in Goheung on 21 October 2021.
Source: Picture Alliance / AP / Korea Pool / Yonhap

Naro 1 being unloaded at Gimhae International Airport, 450 km
southeast of Seoul (2012). Jointly made by the Republic of Korea
and Russia, it made its way to the space centre by plane.
Source: South Korean Science Ministry

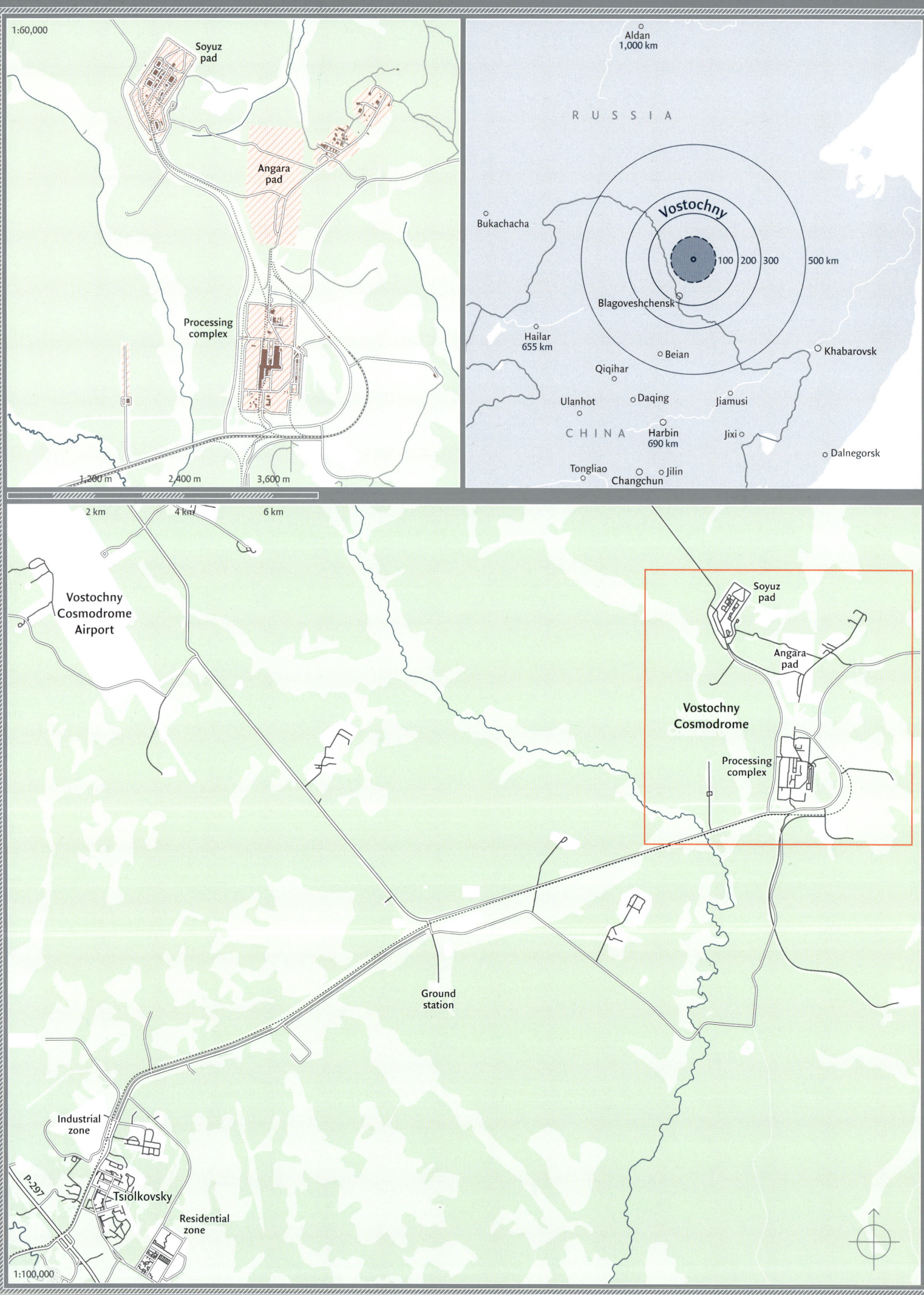

1:60,000

Soyuz
pad

Angara
pad

Processing
complex

1,200 m   2,400 m   3,600 m

Aldan
1,000 km

RUSSIA

Vostochny

Bukachacha

Blagoveshchensk

100 200 300   500 km

Hailar
655 km

Beian

Khabarovsk

Qiqihar

Ulanhot   Daqing   Jiamusi

CHINA   Harbin   Jixi
690 km

Tongliao   Jilin   Dalnegorsk
Changchun

2 km   4 km   6 km

Vostochny
Cosmodrome
Airport

Soyuz
pad

Angara
pad

Vostochny
Cosmodrome

Processing
complex

Ground
station

Industrial
zone

P-297

Tsiolkovsky

Residential
zone

1:100,000

NEW MODEL COSMODROME IN RUSSIAN FAR EAST

Name
Vostochny Cosmodrome

Location
Vostochny, Russia

Owner/Operator
Roscosmos State
Corporation

Elevation
260 m

Coordinates
51.5°N, 128.3°E

Time zone
GMT+9

Launches
290 satellites

Completion
2016

Vostochny is Russia's newest showcase cosmodrome and both technically and architecturally one of the most advanced in the world. Meaning 'eastern', it is located in the far east, near the Amur River. The idea of a launch site here dates to 1993. During the breakup of the Soviet Union, Baikonur cosmodrome became landlocked inside Kazakhstan, with Russia paying rent and depending on the goodwill of the Kazakh authorities for its continued operation. Accordingly, Russia explored the idea of an eastern cosmodrome at a latitude similar to Baikonur (51°). Although distant, it could use the Trans-Siberian railway and launch out to sea over the Pacific, where good radar and tracking facilities were already in place. Initially, a new cosmodrome was opened in nearby Svobodny, an old missile base, with five satellites launched using converted missiles between 1997 and when it was closed in 2006. There was a

small town for missile personnel nearby, Uglegorsk ('coal town') – the name a deception to convince the unsuspecting or the merely unquestioning that it was really mining town.

The following year, 2007, a large cosmodrome on a new nearby greenfield site was announced, both as a long-term replacement for Baikonur and as a significant civil project for high-tech regional investment. Up to 30,000 people were expected to live nearby. It was the first big infrastructure project in the post-Soviet Russian space programme, which had been hitherto starved of funds for research, development, infrastructure, and even cyclical maintenance.

Construction of Vostochny has been peristaltic, as unreliable space budgets have flowed unevenly. The project has been dogged by corruption, leading to as many as 80 prosecutions,

A launch site from scratch: In historic fashion, Russia shows its expertise in construction in the remote eastern parts of the country.
Source: Picture Alliance / dpa / TASS / Sergei Savostyanov

The remoteness of the launch pad raises not just technological infrastructural challenges, but also a need for adequate workers' housing.
Source: Dima Ostrovsky / Flickr.com

accompanied by cost overruns, ramp-ups ahead of inspection visits, and multiple problems with contractors. It was not until 2012 that clearing began for the first launch pad, which was designated for the Soyuz 2 launcher. A 13 km railway connection to the Trans-Siberian railway at Promyshlennya was the first stage to be built. Work repeatedly fell behind schedule and at one stage, 1,200 student volunteers were drafted during mid-winter, echoing the work brigades of the Soviet era. Workers went unpaid for months. There were demonstrations and strikes – even a hunger strike – poor conditions, barrack housing, health and safety violations, and a lack of clothing to cope with -40°C temperatures.

The Soyuz rocket required a substantial amount of digging for its large flame trench and needed a huge metal structure called the tulipan to hold the rocket on the pad, followed by the installation of the large underground service cabin and complex machinery. In a departure from the practice at Baikonur and Plesetsk cosmodromes, Vostochny uses a vertical structure for the rocket's final assembly, called the mobile service tower, as in Kourou (qv). This protects the rocket from the elements until it is ready for launch, enabling it to by readied by technicians up to 37 m above the ground in enclosed conditions. The mobile tower itself is 48 m high, brings the rocket about 100 m down to the launch pad, and then rolls back about an hour before take-off. The tower is painted in various shades of blue, with white stripes, sometimes adorned with slogans. It is a dominant feature of the Vostochny skyline. Also visible beside the pad are two high red-and-white painted lightning towers.

Although the launch pads have been the prime focus of construction at Vostochny and the most common metric of its progress,

at least as important was the 104-hectare Technical Processing (*Tekhnicheskaya Pozitsiya*) area – a massive complex of vehicle, spacecraft, storage, and power supply buildings, principal of which was the large spacecraft processing building, the MIK KA. The 8,640 m² MIK KA has seven floors, topped by two 50-tonne cranes 34 m above the floor. Shipped from St Petersburg, the span is 42 m, designed to service the Soyuz and its various upper stages, such as the Briz, Fregat, and Volga. The other main buildings are the rocket assembly building (MIK RN), a transborder gallery to move equipment between buildings, the fuelling station (ZNS), a power block, as well as fuel production and storage facilities (oxygen, nitrogen, kerosene, napthyl). The MIK RN has an area of 9,000 m² and can handle two complete Soyuz rockets simultaneously, moving along four railway lines on the floor, lifted by two cranes measuring 27 m in height and 40 m across, each able to hold 100 tonnes. The facilities are not only climate-controlled but operate to high cleanroom standards. The windows of the complex alone let in 13,000 m² of light. Finally, 450 m from the pad is a new 3,000-m², two-floor launch control centre, blast-proof to survive a nuclear-level pad explosion. Vostochny, alongside the Chinese launch side in Hainan, represents the most modern thinking in cosmodrome design worldwide. The first launch finally took place on 28 April 2016, putting the Lomonosov scientific satellite into orbit. So far, the launch rate has been slow. It was followed by two weather satellites (Meteor 2-1, failure in 2017, and Meteor 2-2 in 2019), two Earth observation satellites (Kanopus V 3 and 4; Kanopus V 4 and 5, both 2018), all carrying micro-satellites as additional payloads.

Almost 60 years of launch site experience informs Russia's latest remote concrete pads. What looks like a busy highway is the meticulously planned infrastructure around Russia's model launch site.
Source: Tsenki Vostochny Space Centre

Left:
The recessed launch pad and gantry arm set-up for Soyuz rockets is quite different from both European and American practice.
Source: Picture Alliance / dpa / Igor Ageyenko

Right:
2021: A Soyuz-2.1 b is transported out of the assembly area horizontally to be hoisted onto the gantry for launch.
Source: Roscosmos Press Office

Next double page:
Despite some launches, the Vostochny Cosmodrome has yet to show much aging, unlike the much older Baikonur site with its distinct rugged appearance.
Source: Picture Alliance / dpa / TASS / Valery Sharifulin

It was always the plan for Vostochny to house new-generation rockets, so work then proceeded to build a second pad, this time for the new Angara rocket. The rocket was intended to replace the Proton designed in the 1960s. Angara's heaviest version is able to lift up to 25 tonnes into orbit, with other versions for lighter payloads. In particular, a version of Angara may lift the new manned spacecraft to replace the 1960s Soyuz, the PTK-NP, later named the Federatsiya, and finally the Orel.

Construction of the Angara pad has been uneven, with completion scheduled for 2022 with the first Angara launch from here in 2023. The earlier difficulties now seem well in the past.

Work is due to begin on the third launch pad 2025. It is intended for the medium-lift Soyuz 5 rocket. This is to be followed by a very large rocket, the Yenisiei, intended to enable Russia to match the American Space Launch System and China's Long March CZ-9 and send cosmonauts to land on the Moon. Vostochny appears to be one of the big 'grands projects' of Vladimir Putin as President and of Dmitry Rogozin, his deputy prime minister responsible for the space industry since 2011 and head of Roscosmos (2018-2022). Rogozin was very hand-on with the project and visited Vostochny on numerous occasions to troubleshoot, encourage, admonish, and speed-up progress.

Vostochny has a picturesque location, surrounded by evergreen conifer forests that stretch to the horizon, giving it the appearance of an airy location with clear light and abundant sunshine. It has been given prominence in Roscosmos publicity. Unlike some other parts of the space programme, everything here looks new – it is new – and it is literally a very concrete way of showing that renewal

in the programme is really underway. Rockets and spacecraft travel to Vostochny from Moscow and Samara – where the Soyuz rocket is made – on the Trans-Siberian railway, whose extension runs right into the buildings, hauled into them by big powerful diesel engines. A passenger railway station, Ledyanaya, was also completed. Further away from the pads, a tracking station was built on the disused Svobodny missile base. Although Svobodny has a small airport 30 km away and a larger one at Blagoveschensk 180 km away, a nearer airport will be built, especially to take outsize cargoes too big to travel by rail and able to take the big Antonov freighters. The first clearing began in 2015 but progress has been slow. In line with Soviet era planning practice, housing was to be constructed simultaneously, but was so slow that it was berated by President Putin during a visit. Rather than use the small, largely old town of Uglegorsk, whose apartments emptied after the Svobodny base closed, it was decided to build a new city called Tsiolkovsky, after Konstantin Tsiolkovsky (1857–1935), the father of cosmonautics. Tsiolkovsky follows the traditional Soviet design of apartments, but with more room, light, space, and design features. Siberian homes have always been better than those of the big cities – part of the process of incentivising workers to come there – along with improved retail facilities and food. For Tsiolkovsky, there were six-, nine-, and 12-floor apartment blocks with wide pavements, cycleways, ramps, and disability access as well as kindergartens, recreational facilities (e.g., a swimming pool), a hospital, a children's medical centre, shops, a bus route, an administrative centre with conference rooms, and even a marriage office. The first residents moved into their new apartments in 2016.

Top:
Some might mistake the scene for an unfinished stadium. Vostochny's new launch pad for Angara rockets is scheduled to be completed for its first launch in 2023, followed by the first crewed spacecraft two years later.
Source: Picture alliance / dpa / TASS / Alexander Ryumin

Bottom:
The manual labour involved in constructing the massive launch sites is an often-overlooked human factor in space exploration and stands in stark contrast to the pristine white rocket assembly halls.
Source: Picture Alliance / dpa / TASS / Sergei Savostyanov

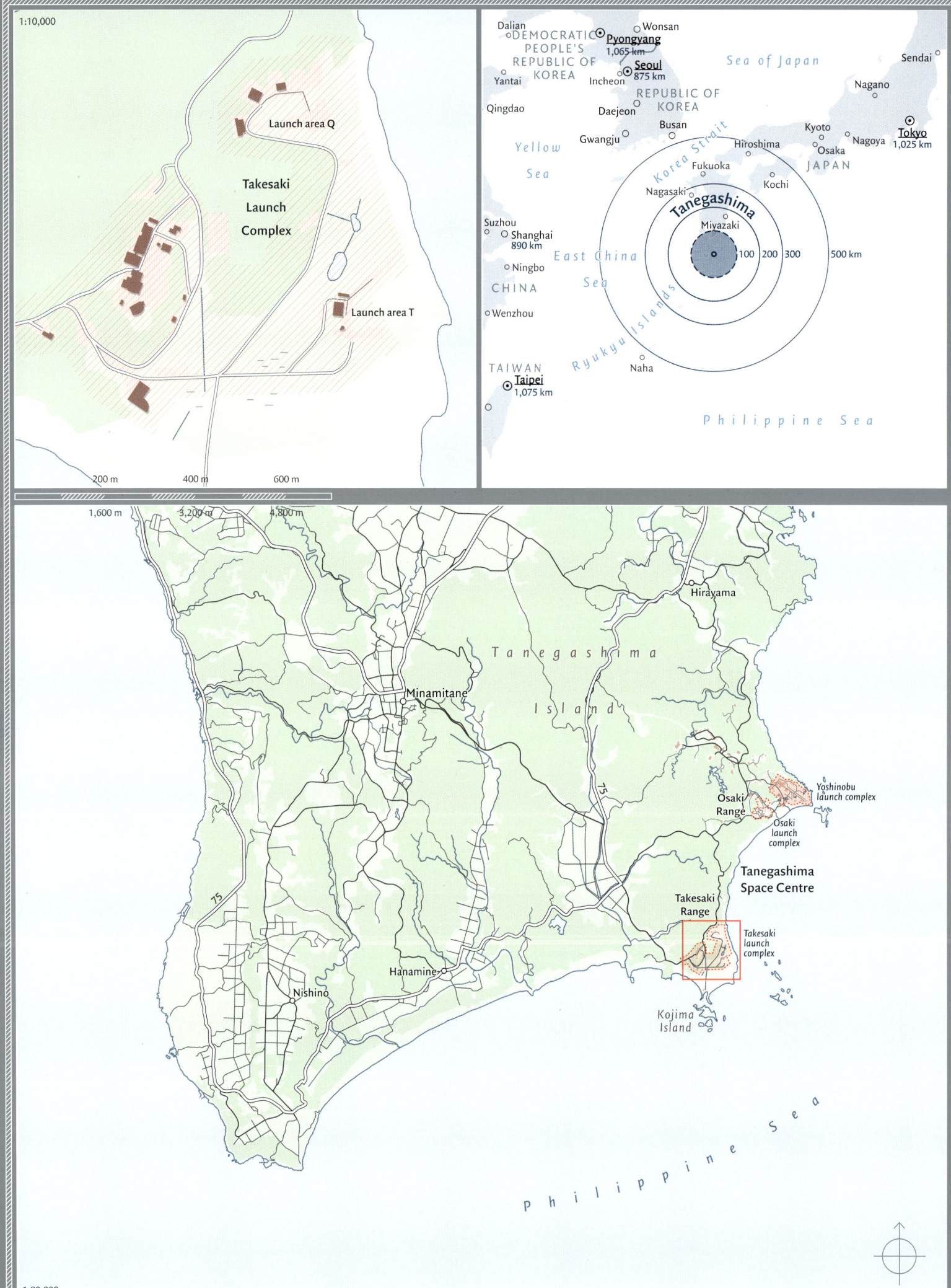

1:10,000

Launch area Q

Takesaki
Launch
Complex

Launch area T

200 m   400 m   600 m

Dalian
DEMOCRATIC   ⊙ Pyongyang
PEOPLE'S   1,065 km
REPUBLIC
OF KOREA   Sea of Japan
Yantai   Incheon ⊙ Seoul   Sendai
                       875 km
Qingdao   REPUBLIC OF   Nagano
             KOREA
Gwangju Daejeon   Kyoto   ⊙ Tokyo
             Busan   Osaka Nagoya   1,025 km
Yellow   Hiroshima
Sea   Korea Strait   Fukuoka   Kochi JAPAN
                                         JAPAN
Suzhou   Nagasaki
                      Tanegashima
Shanghai
890 km   East China   Miyazaki
             Sea
Ningbo   100 200 300   500 km
CHINA

Wenzhou   Ryukyu Islands

                        Naha
TAIWAN
⊙ Taipei
1,075 km   Philippine Sea

1,600 m   3,200 m   4,800 m

                              Hirayama

                   T a n e g a s h i m a

                        I s l a n d

        Minamitane
                                        Yoshinobu
                                        launch complex
                              Osaki
                              Range   Osaki
                                        launch
                                        complex

                              Tanegashima
                              Space Centre

                      Takesaki
                      Range   Takesaki
                                 launch
                                 complex
Hanamine

Nishino

                Kojima
                Island

                    P h i l i p p i n e   S e a

1:80,000

ISLAND LAUNCH SITE AT JAPAN'S SOUTHERN TIP

Name	Tanegashima Space Centre	**Coordinates**	30°N, 130°E
Location	Minamitane, Japan	**Time zone**	GMT+9
Owner/Operator	Japan Aerospace Exploration Agency	**Launches**	145 satellites
Elevation	20 m	**Completion**	1975

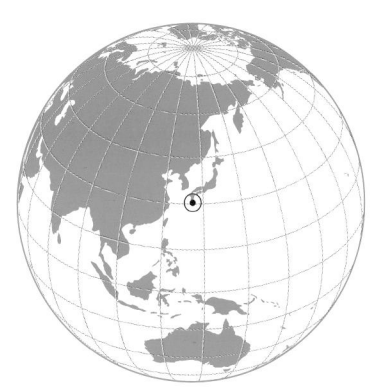

The Tanegashima launch site is on a small 8.6-km² island off the very southern end of the Japanese archipelago, a picturesque coastal site, set in rolling hills alongside a rocky-sandy seashore where gentle waves roll in from the Pacific. It is not unlike the smaller Uchinoura launch site (qv) 100 km away. Likewise, it is a relatively under-populated area and had the advantage of all of the Pacific Ocean over which to launch. It had the same disadvantage of having to negotiate launch dates with local fishermen. But why two launch bases so relatively near to one another? The Uchinoura base was the home of Hideo Itokawa's Institute of Space and Aeronautical Science of the University of Tokyo, the scientific space programme. In 1969, under pressure from industry, the government approved a national space programme and set up the National Space Development Agency (NASDA). The

two later merged as the Japanese Aerospace Exploration Agency (JAXA) in 2003. The national space programme had its own programme, launchers, facilities, and launch site.

The original launch site, which cost ¥1.466bn (€13m) to build in the 1970s, was constructed on an older 1960s Takesaki pad for sounding rockets at the southern end of the bay, which was extended for tracking and solid rocket engine tests, and a new site, called the Osaki complex, at the northern end, with a pad, a high mobile service tower, propellant storage facilities, a solid motor test building, a spin test building, clean rooms, a third stage and spacecraft assembly building, as well as a semi-underground control building. A power plant, range control centre, telemetry station, and checkout facility were built alongside. The now-deactivated Osaki complex served Japan's first generation of launchers, the

N series, a derivative of the American Thor rocket, built under licence and American supervision. The rocket was prepared in a mobile service tower before being rolled back 100 m before launch. For the final stages of the countdown, the rocket was serviced by two umbilical towers. The first N rocket, Tanegashima's first orbital launch, was launched from here on 9 September 1975. It was the Engineering Test Satellite 1 (ETS 1), also called Kiku 1, and Japan became the third country to reach 24-hour, geosynchronous orbit. The N rocket was used thereafter to launch a broad range of Japanese communications, weather, engineering, test, and applications satellites in the 1990s, both to low Earth orbit and geosynchronous orbit. The more powerful N-II rocket, a derivative of the American Thor Delta, flew from Tanegashima's Osaki site from 1981–1986. Osaki's last launch was the J-I solid rocket booster, but it only flew once during the HYFLEX space plane test mission in 1996. NASDA then introduced its H series of rockets, with a powerful hydrogen-fuelled upper stage on the H-I and both stages from the H-II and its successors – the H-IIA, H-IIB, and then the H-III. These rockets were ever more indigenous (the H-I 80 per cent so, the H-II wholly so). For them, the purpose-built Yoshinobu complex was constructed one kilometre away on the headland at the northern end of the bay on an area covering 150,000 m². It was not ready until the H-II, so the H-I, first launched on 13 August 1986, continued to use the Osaki site and from there launched a series of communication, weather, and Earth observation satellites over the next six years until 1992. Anticipating a higher launch rate, Yoshinobu was a double pad able to handle two large rockets at a time. This is a much larger and impressive site, with the first pad completed in 1992 and the second in 1999. Other facilities were built close by, including a static firing test centre for liquid fuel engines, enabling the first stage to be held down on the pad to fire the H rocket's LE-7 engines, as well as a big cement deflector trench, a water coolant system, and, nearby, liquid hydrogen and liquid oxygen storage tanks, and a high pressure gas storage facility for helium and nitrogen. The Yoshinobu site comprises a vehicle assembly building with a mobile launcher, pad service tower, control centre, and cryogenic storage services. The vehicle assembly building, in appearance not unlike the Cape Canaveral one, measures 81 m in height, 64 m in width, and 34.5 m in depth. Rockets are brought the short distance to the pad on the mobile launcher, which rolls down an apron with parallel roads on either side and a grass centreline with the large letters of the space agency carved into the greenery. Spacecraft and rockets are assembled vertically on a mobile tower in the vehicle assembly building, which has a number of bays. The tower is 22 m high, weighs 800 tonnes, and once everything is ready, makes the 500-metre rail journey down to the pad. The crawler has 56 wheels on 14 axles, travels at 2 km/hour, and takes 25 minutes to reach the nearby pad. Here the rocket meets a pad service tower – a fixed section with two rotating parts. This is 67 m tall, with 12 floors, and uses a 20-tonne crane. Launches at Tanegashima are controlled from an underground blockhouse 170 m away, entered through a hexagonal, white, tent-shaped building some 500 m from the pad, with controllers taking a lift downward. Because the H-IIA carried 275 tonnes of hydrogen fuel, any explosion was likely to be a large one, so the centre was ten floors underground with a

The launch centre itself is planned in a
very compact and efficient manner, fitting
in tightly with its cliff face surroundings.
Source: JAXA

Previous double page:
Tanegashima Space Centre overlooks a
picturesque beach on the Philippine Sea.
Source: JAXA

The Tanegashima Space Centre (top) to the south of the actual launch
complex houses both the control centre and an observation stand (bottom),
which includes the on-site space museum.
Source: JAXA

Top:
One of the boosters being hoisted onto N-II launch vehicle No.5 at the vehicle assembly building.
Source: JAXA

Middle:
N-II Launch Vehicle No.8 is ready to be launched from the beachside pad in February 1984.
Source: JAXA

Bottom:
The second stage is hoisted to the top of N-II Launch Vehicle No.8 in the evening hours.
Source: JAXA

Right:
The H-IIB rocket No. 5 makes the short journey from the vehicle assembly area to the launch pad under the cover of darkness.
Source: JAXA

1.2-metre-thick ceiling, but it is bright and airy. The air conditioning system can provide air for one hundred controllers for four hours (there is an evacuation tunnel in any case). Yoshinobu was inaugurated on 4 February 1994 with the H-II, considered one of the most sophisticated rockets in the world when it was introduced. It was followed by the H-IIA on 29 August 2001. Both were technically challenging. Their introductions had their share of difficulties, but they eventually achieved a steady rate of launches from Tanegashima of Earth observation, weather, military intelligence, and technology testing satellites. Japan was one of the partner nations on the International Space Station (ISS), contributing one of the space station's modules, Kibo, brought up by the space shuttle, as well as the large cargo vessel, the 16-tonne HOPE Transfer Vehicle (HTV). This required a more powerful launch vehicle, the H-IIB, based on the H-IIA but with a larger first stage and four solid fuel rockets attached. The first H-IIB, with the first HTV, was launched from Tanegashima on 10 September 2009. Long-range tracking cameras meant that it was possible to follow the whole first stage burn and subsequent coasting far into the Pacific sky downrange. The Yoshinobu complex, with the H-II and HTV, gave Japan top-level access to the ISS, with its astronauts flown there by Russia and the United States.

The H rocket did indeed achieve the higher launch rate intended, so new agreements were reached between the government and the fishermen's union to extend the launch window from 90 days to 180 days, while still preserving the prime fishing period of March to mid-June, with exceptions for rockets that can only be launched at particular times. Fishing restrictions were finally lifted in April 2011. The H-IIA is used from the first pad and the H-IIB from the second, with the latter now being converted for the next version of the rocket, the H-III, in the 2020s.

The Tanegashima facility covers 9.7 km² and includes a tracking and communications station, two radar stations, an 80-metre-high meteorological tower, and an optical observation post. The original Takesaki site is still used for sounding rockets. There is a museum for visitors, with a full-scale model of Kibo and the Daichi satellite, and visitors can experience a rocket launch 'with full audio'.

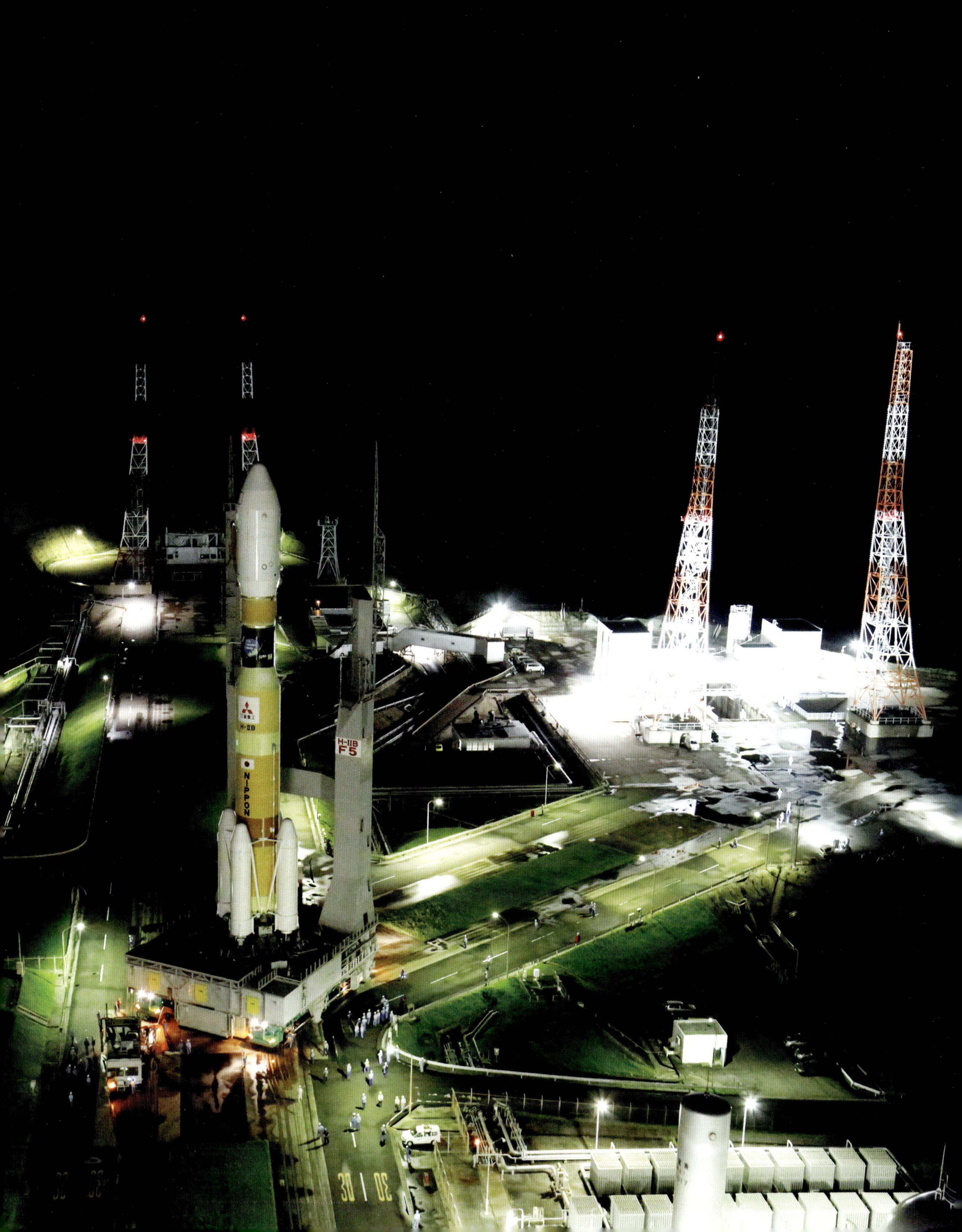
NIPPON
H-IIB
F5

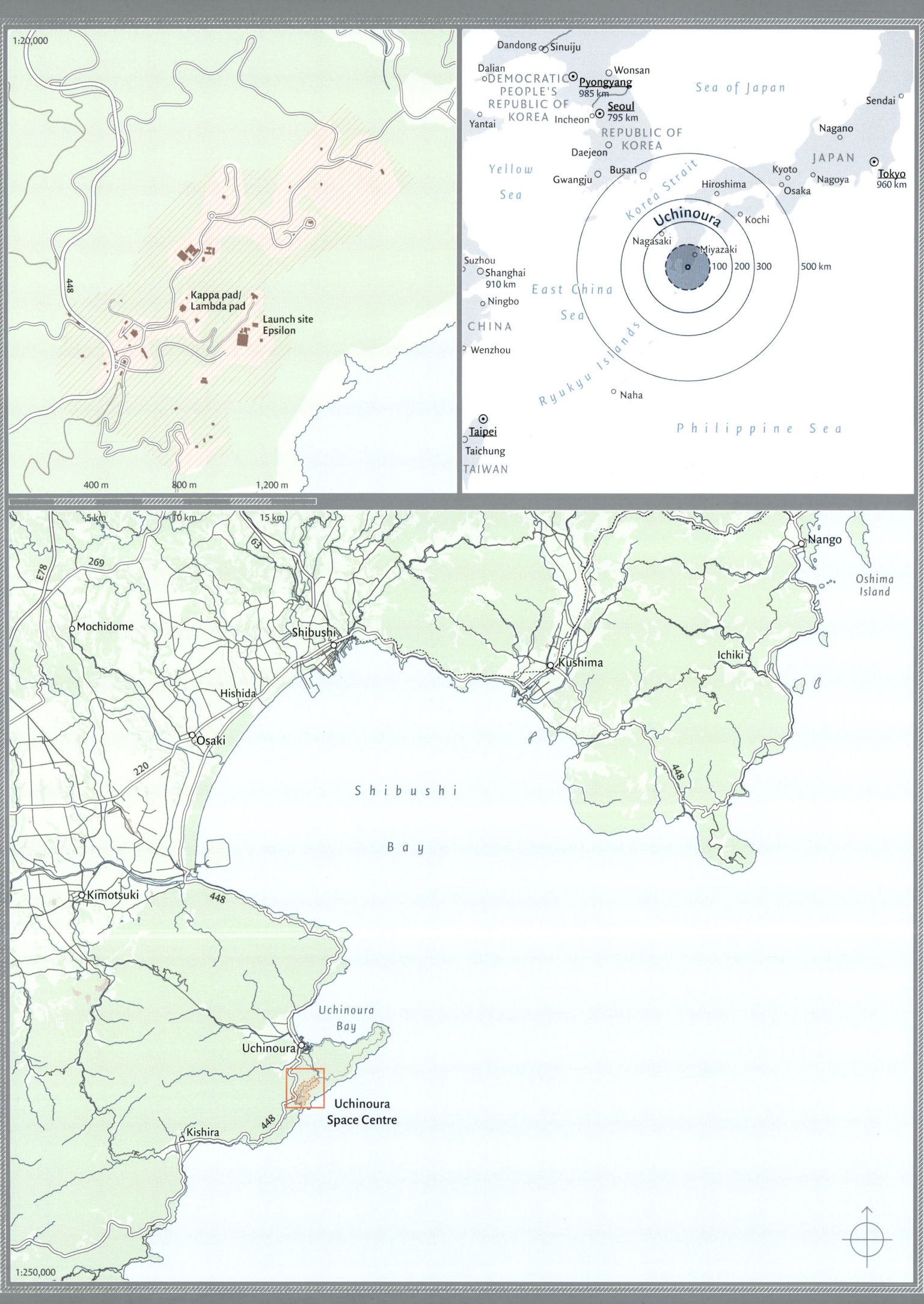

1:20,000

Kappa pad/
Lambda pad

Launch site
Epsilon

448

400 m   800 m   1,200 m

Dandong ○ Sinuiju
Dalian
DEMOCRATIC
PEOPLE'S
REPUBLIC OF
KOREA
Yantai   Incheon ◉ Seoul
Daejeon   795 km

○ Wonsan
◉ Pyongyang
985 km

Sea of Japan

Sendai

Nagano

REPUBLIC OF
KOREA
Gwangju   Busan
Korea Strait
Hiroshima   Kyoto
Nagoya
Osaka

JAPAN

◉ Tokyo
960 km

Yellow
Sea

Nagasaki
Uchinoura
Miyazaki

Kochi

100 200 300   500 km

Suzhou
Shanghai
910 km
○ Ningbo
CHINA
Wenzhou

East China
Sea

Ryukyu Islands

○ Naha

Philippine Sea

◉ Taipei
Taichung
TAIWAN

5 km   10 km   15 km

E78   269
63

Nango

Oshima
Island

Mochidome

Shibushi

Kushima

Ichiki

Hishida

Osaki

220

Shibushi

Bay

448

Kimotsuki   448

Uchinoura
Bay

Uchinoura

Kishira   448

Uchinoura
Space Centre

1:250,000

KAGOSHIMA: JAPAN'S FIRST LAUNCH SITE

Name
Uchinoura Space Centre

Location
Kagoshima, Japan

Owner / Operator
Japan Aerospace
Exploration Agency

Elevation
210 m

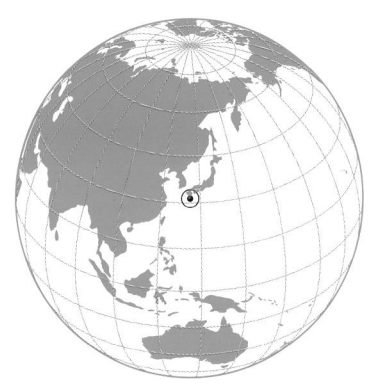

Coordinates
31.4°N, 131.1°E

Time zone
GMT+9

Launches
48 satellites

Completion
1970

Uchinoura launch site in Kagoshima was Japan's first orbital launch site, but over time became less used than the bigger site in Tanegashima (qv) of the National Space Development Agency (NASDA). The chief designer of the Japanese space programme, Hideo Itokawa, had first fired tiny rockets from populated beachside sites, but in 1959 searched for a safer, less populated site on the Pacific with a good downrange area. Seven were considered: Uchinoura, at the tip of Japan's southernmost island, subtropical Kyushu, was selected, although its 510-hectare site was far from Tokyo, had hilly tree-covered rocky terrain, and launch timings had to be agreed with local fishermen. Uchinoura was developed by the Institute of Space and Aeronautical Science (ISAS; later 'Astronautical'), the scientific part of the Japanese space programme and used small solid-fuel rockets.

Construction started in 1962: assembly, payload handling, integration and administrative buildings, and a control centre called the Lambda centre. Sounding rockets were launched from here in 1964, fired from an enclosure with rails to guide them out over the sea at an 80° angle. The first orbital attempt using a small solid-fuel rocket failed in 1966, with success later being achieved on the fourth attempt on 11 February 1970. A small scientific satellite, Ohsumi, was put into orbit, making Japan the fourth nation with the capability to build, launch, and operate its own satellites.

A second, larger facility, the Mu Centre, comprising an all-up steel launch tower, assembly building, control tower, and propellant area, was built in 1966. This launched the Mu rocket, which put a series of small scientific satellites into orbit from the 1970s onward. Mu was used in 1985 to send Sakigake and Suisei to the

In true pencil rocket fashion, this launch vehicle
is used to propel a microsatellite into orbit.
Source: Picture Alliance / AP Images /
Daisuke Urakami

The early days of Japanese space exploration did not
feature the sterile procedures used today. A pencil
rocket is loaded onto the launcher in the sand.
Source: JAXA

A Kappa 8 type rocket on a mobile launcher
angled for lift-off. The gantry arm is still a
familiar sight in Japanese launches.
Source: JAXA

Halley comet and the Moon probes Hagoromo and Hiten in 1990,
making Japan the third country to fly deep space and lunar missions. The Mu pad was redeveloped for the more powerful Mu-5
launcher in 1997. Japan's first interplanetary probe, Nozomi, was
launched from here to Mars in July 1998. In 2003, ISAS and NASDA
were merged to form the Japanese Aerospace Exploration Agency
(JAXA), so JAXA took over Uchinoura. The Mu rocket was retired in 2006, so there were no launches until the new solid-fuel
Epsilon rocket first flew in September 2013, starting a modest current manifest of about a launch a year. Kagoshima is Japan's first
and most famous launch site, with a spectacular location.

Right:
Looking like an upscaled version of
its predecessors, 2019 saw the fourth
Epsilon launch vehicle lift-off from JAXA's
Uchinoura Space Centre.
Source: JAXA

Prelaunch spectacle: The unveiling
of JAXA's then new Epsilon solid-fuel
rocket in 2013.
Source: Picture Alliance / Kyodo / *

Right:
People watch the new rocket's launch
from the nearby town of Kimotsuki.
Source: Picture Alliance / Kyodo / *

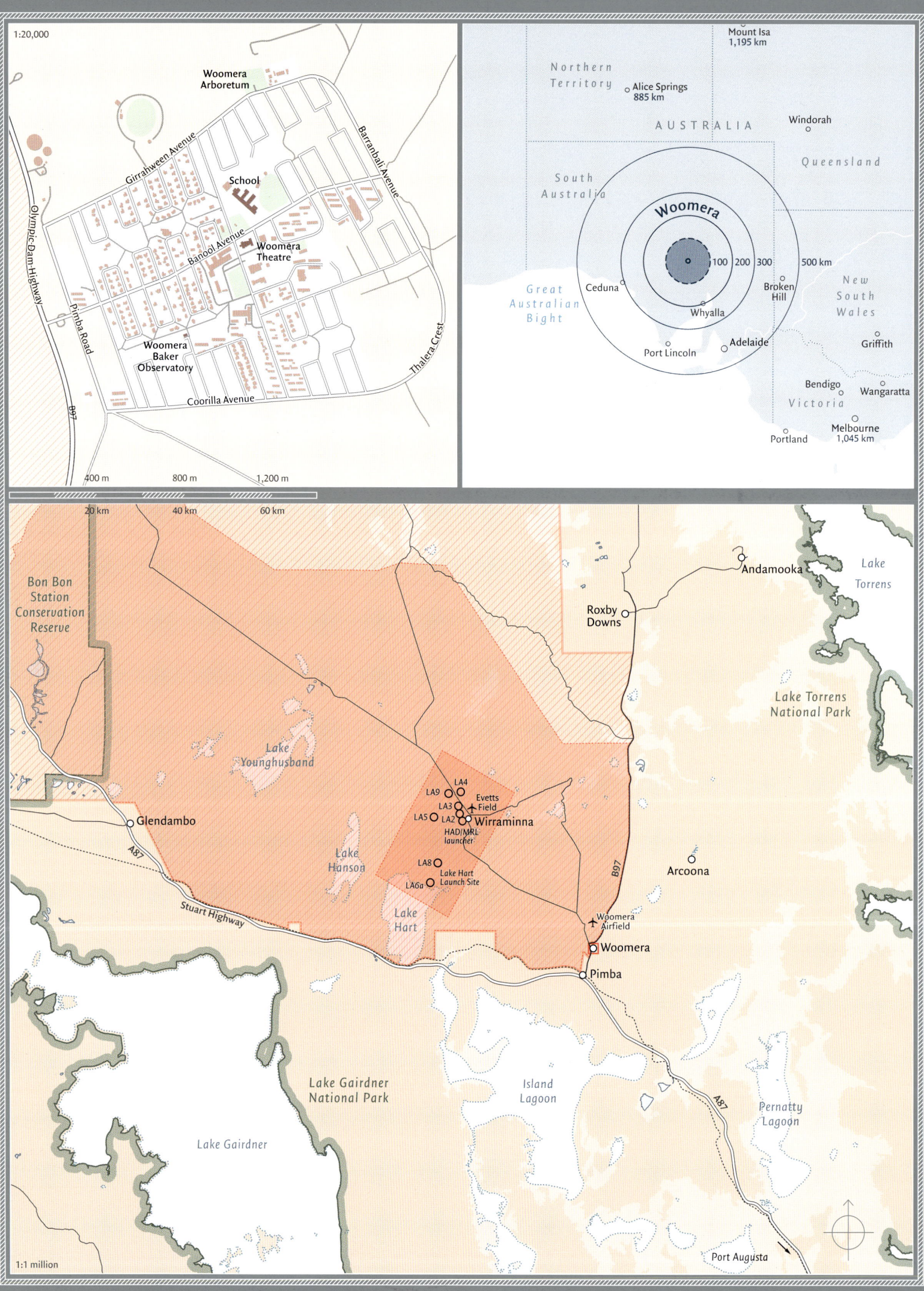

1:20,000

Woomera Arboretum

Girrahween Avenue

Olympic Dam Highway

Pimba Road

B97

School

Banool Avenue

Barranbali Avenue

Woomera Theatre

Thalera Crest

Woomera Baker Observatory

Coorilla Avenue

400 m 800 m 1,200 m

Northern Territory

Mount Isa 1,195 km

Alice Springs 885 km

AUSTRALIA

Windorah

Queensland

South Australia

Woomera

100 200 300 500 km

Great Australian Bight

Ceduna

Whyalla

Broken Hill

New South Wales

Port Lincoln

Adelaide

Griffith

Bendigo

Wangaratta

Victoria

Portland

Melbourne 1,045 km

---

20 km 40 km 60 km

Bon Bon Station Conservation Reserve

Lake Younghusband

Andamooka

Lake Torrens

Roxby Downs

Lake Torrens National Park

Glendambo

A87

Lake Hanson

LA9
LA4
LA3
LA5 LA2
Evetts Field
Wirraminna
HAD/MRL launcher

Stuart Highway

LA8
LA6a
Lake Hart Launch Site

Lake Hart

B97

Arcoona

Woomera Airfield

Woomera

Pimba

Lake Gairdner National Park

Island Lagoon

Lake Gairdner

A87

Pernatty Lagoon

Port Augusta

1:1 million

DESERT LAUNCH SITE IN THE LAND DOWN UNDER

Name
RAAF Woomera
Test Range

Location
Australia,
Oceania

Owner / Operator
Royal Australian
Air Force

Airport type
Military & Civil aerospace

Elevation
169 m

Coordinates
31.1°S, 136.5°E

Time zone
GMT+9:30

Launches
2 satellites

Completion
1967

It had long been axiomatic that Britain would be the European leader in post-war rocketry. The British Isles had the disadvantage of cloudy weather and a lack of suitable trajectories over unpopulated areas, so in 1946, a location was chosen in the south Australian desert 430 km from Adelaide near the small town of Woomera (meaning 'spear-throwing device'). The site was managed by the military Weapons Research Establishment (WRE) of the two countries. The original site comprised an airfield, a village 8 km away, a launch site 40 km to the northeast by utility cars, whose main challenge was to avoid accidentally hitting kangaroos en route, and distant overland tracking sites (e.g., Talgarno). Being a military establishment, it was run on semi-military lines, with messes and wartime-style accommodation where scientists were billeted. The Woomera range extended an enormous 127,000 km² across the desert, which is lightly populated with Aboriginal people and sheep farmers and offers clear skies almost all year. At its 1960s peak, 4,500 people lived here in 500 homes for military and civilian workers. A 2,375-metre runway was built and there were seven designated launch areas. Rockets were fired either northward or westward over observation posts towards the Indian Ocean. Woomera was best known internationally as the launch site for the European Launcher Development Organisation (ELDO) launcher Europa, where Britain made the first stage, the Blue Streak. ELDO's Europa I was fired from Woomera five times (sub orbit) between 1964 and 1966 and five times between 1967 and 1970 (orbital attempts), but none were successful and the successor launcher, Europa II, was moved to Kourou (qv). Britain's small Black Arrow was launched from

The southern launch pad overlooking Lake Hart salt lake
is reminiscent of a James Bond villain's base.
Source: Australian National Archive

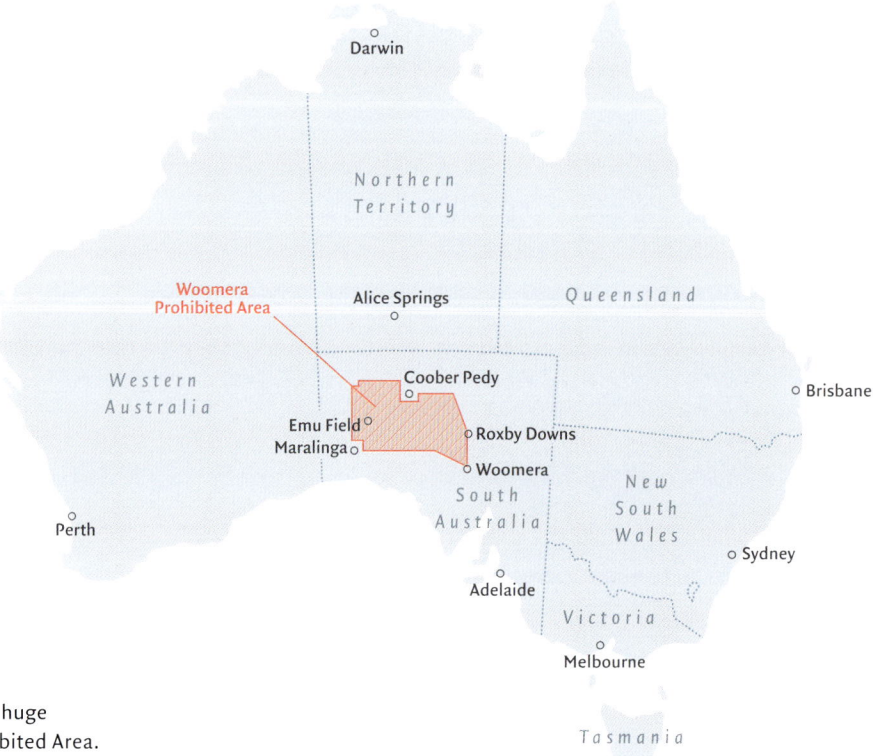

Map of Australia indicating the huge
territory of the Woomera Prohibited Area.
Source: Paul Meuser

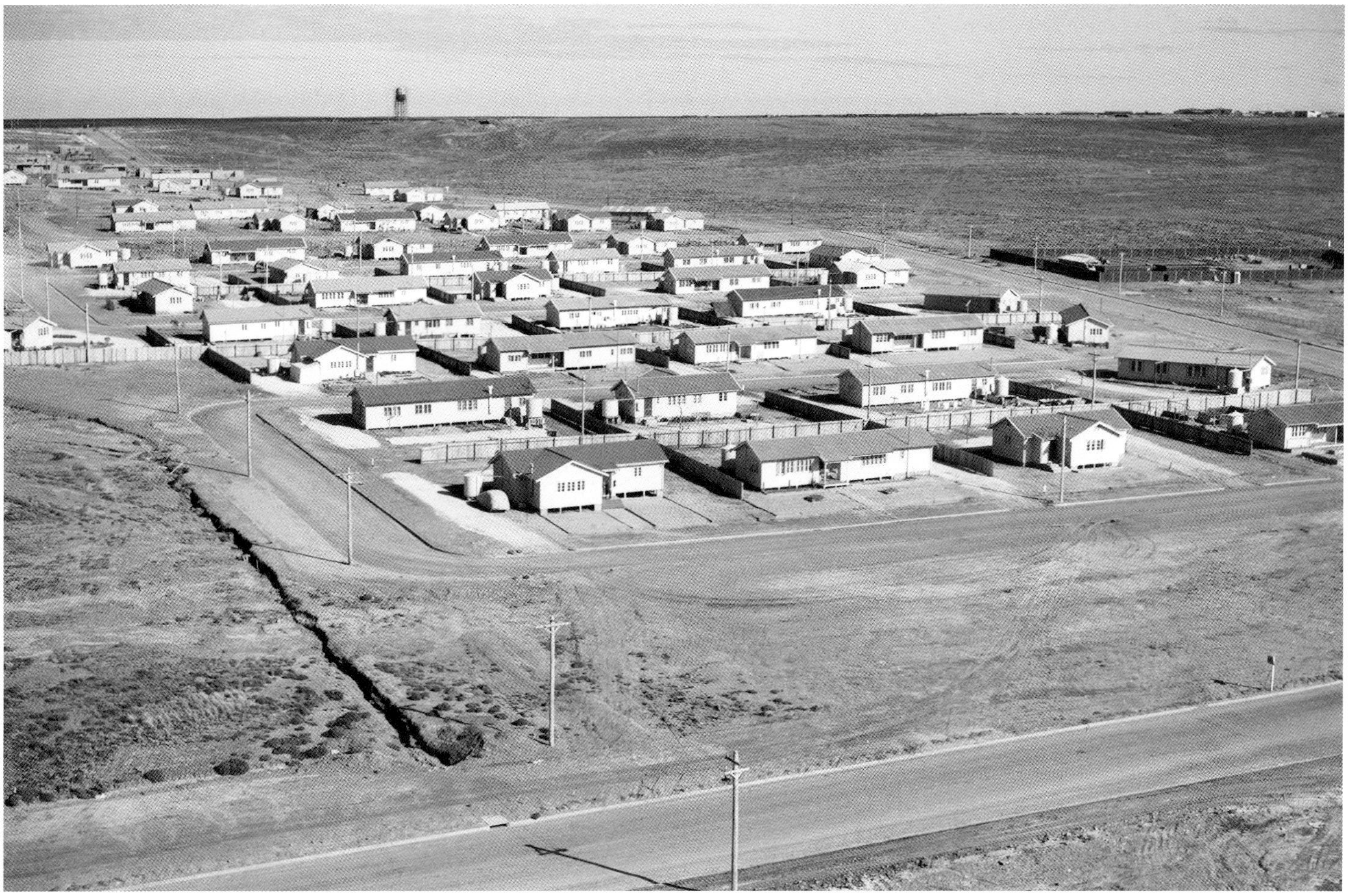

Radioactive warning sign outside the Maralinga test site 500 km west of the town of Woomera, marking the western corner of the overall prohibited area. Overall, there were 12 major nuclear weapons tests conducted in Australia.
Source: Australian National Archive

War and amenities: To accommodate the scientific and military personnel, the township of Woomera provided space for up to 6,000 people at its height in the 1960s.

Source: Australian National Archive

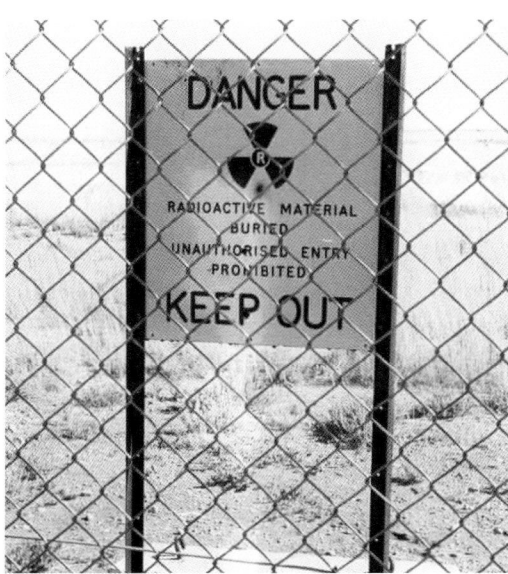

Woomera on two suborbital and two orbital missions, the final one being successful, orbiting Prospero in 1971. An Australian satellite, WRESAT, was also put into orbit from here on the American Redstone in 1967. Woomera was the base for Britain's successful Skylark sounding rocket from 1957 to 1979, then comprising an integration and test shop, launch tower with rails, and an underground blockhouse. Woomera has since been used for occasional sounding rockets, to test Japan's ALFLEX space plane (1996), to recover Japan's Hayabusa asteroid explorer (2010), and as a testing ground for solar-powered aircraft (2020). Woomera is largely abandoned, with a small population of 130 to 150. It has a souvenir shop, museum, and a graveyard of rocket parts. The officers' mess, later the ELDO hotel, still stands.

Officers during one of the first parties thrown outside their still makeshift-looking mess in 1950. It was six years before the first nuclear test on the site and around eight years before the first launch attempt in 1958.
Source: Australian National Archive

Cars parked outside the officers' mess.
Source: Australian National Archive

The appearance of the mess bar is a relic of its time and offers a glimpse into the off-duty time of the base's officers, who were looking for a portion of mundane daily life between the hot desert and the rocket launches.
Source: Australian National Archive

Technicians cramped into the
small control centre.
Source: National Archive Australia

Mobile sensor and control stations
housed in containers.
Source: Australian National Archive

Right:
A lonely outpost standing watch over
the desolate road.

Bottom:
A Woomera pool filled with women
and children.

Left:
A deceptively brightly coloured
Thunderbird surface-to-air missile to
be tested on the Woomera Range.
Source: National Archive of Australia

Left:
A crowd gathers around the flame exhaust
section of the test stand to witness the flame
suppression system in action.
Source: National Archive Australia

Bottom:
The silver tower looming over the steep hillside
north of Lake Hart sticks out of the barren
landscape. Looking outwards onto the dusty
salt lake could make you think you're already
on a distant hostile planet.
Source: National Archive Australia

Top:
Operators working in the control room of the deep space
instrumentation facility at Island Lagoon tracking station.

Right:
US and Australian scientists and delegates gathered around
the large 26-metre satellite dish of the tracking station.
Source: National Archive Australia

1:10,000

Mahia
Launch
Complex

Mahia East Coast Road

Rocket lab
complex 1

Ahuriri Bay

Ahuriri
Point

200 m   400 m   600 m

Pacific

Ocean

Tasman
Sea

Auckland

Mahia

Hamilton   Tauranga

North
Island

Napier

100 200 300   500 km

NEW
ZEALAND

Wellington

Nelson

Blenheim

South
Island

Christchurch
660 km

Queenstown

Dunedin
965 km

10 km   20 km   30 km

Gisborne
District

Hawke's Bay

Region

Frasertown

Wairoa

SH 2

Nūhaka

Mahia
Beach

Mahia
Peninsula

SH 2

Mahia
Launch
Complex

Portland
Island
(Waikawa)

Hawke Bay

Hawke's Bay
Airport

Napier

SH 51

1:500,000

NEW, BUSY NEW ZEALAND MODEL COMMERCIAL SPACEPORT

Name
Rocket Lab Launch Complex 1

Location
Mahia, New Zealand

Owner/Operator
Rocket Lab

Elevation
109 m

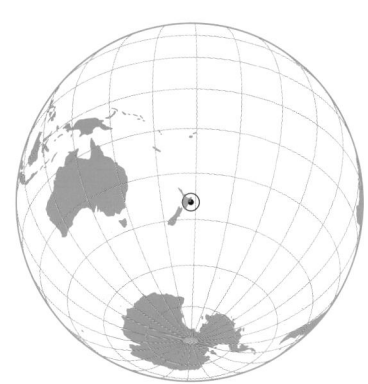

Coordinates
39.3°S, 177.9°E

Time zone
UTC+12:00

Launches
111 satellites

Completion
2018

Mahia, New Zealand, is the world's newest launch site and already one of the busiest. It was set up by Rocket Lab, an American company founded by New Zealanders. Called Mahia Spaceport, it is named after the Mahia peninsular where it is located on the east side of New Zealand's North Island, which is especially suitable for polar orbital flights. It has a coastal cliffside location on Ahuriri Point, Hawke's Bay, and is the first launch site in New Zealand. It is isolated, with the nearest town of Wairoa 50 km away. Limited air and marine activity permit frequent launches.

In the 1990s, the trend towards ever larger satellites was reversed, with micro-electronics making possible smaller, more capable satellites. Originally, they hitched rides on larger rockets, but from the 2010s onwards, commercial companies developed affordable small rockets capable of lifting micro-satellites, typically several at a time. Here, Rocket Lab created the two-stage liquid-fuel

Electron, a slim launcher measuring 17 m in height and 1.2 m in diameter and weighing 12.5 tonnes. It is able to lift 225 kg to 550 km or combinations thereof. The company is able to minimise costs by using high-speed robotic 3D carbon fibre production line manufacturing, with plans to make the rocket recoverable and re-useable.

The Mahia site started with a single launch pad, built in only two years. It is a rectangular-shaped area, with a payload processing facility at the land end leading down to the pad at the cliffside end. The fuel tanks, electrical station, and water deluge tanks are to the left of the pad, while the liquid oxygen facility is on the right. There is a 50-tonne launch tower at the end in the middle. Payload processing includes a hangar, loading bay, two clean rooms with airlocks, and client facilities. There is a control centre and two tracking dishes.

The commercial spaceport saw its
first launch in 2017, with Rocket Lab
not just celebrating its technological
achievements, but also procuring what
is arguably one of the most idyllic
launch sites in the space industry.
Source: Rocket Lab

Right:
Nature vacation retreat or launch site?
Source: Rocket Lab

The first launch attempt was in May 2017 and was unsuccessful,
with orbit later achieved in January 2018. Electron has since made
a further ten launches, all successful, of micro-satellites (now 111)
for both civilian and military customers and has a long order list.
The company has also completed a second pad here and has already
built another at Wallops Island, Virginia (qv). Mahia is a pioneer-
ing privately-funded spaceport, likely be a commercial model for
others. Wairoa District Council gives visitors directions to watch
launches from Bluck's Pit. Bring a picnic, it says.

1:10 million

Stage 1 & 2
powered flight

Block DM-SL burn

SC Separation

Equator

1st Apogee Passage

USA

MEXICO

Pacific

Ocean

Caracas

Quito

PERU

Lima

BRAZIL

Brasília

Atlantic

Ocean

ALGERIA

Abuja

Luanda

Oslo

Reykjavik

Helsinki

Moscow

Greenland

Hudson
Bay

Ottawa

Washington

Nur-Sultan

Arctic

Ocean

CANADA

RUSSIA

Bishkek

Rocky Mountains

USA

MONGOLIA

New Delhi

Ulaanbaatar

Bering
Sea

Kathmandu

MEXICO

Thimphu

Beijing

Los Angeles

Mexico City

Dhaka

Vladivostok

180°

Pyongyang
Seoul

CHINA

JAPAN

Hanoi
Vientiane

Tokyo

Bangkok

Taipei

Northern

Phnom Penh

Kuala
Lumpur

Manila

Pacific Ocean

Hawaii

Singapore

Bandar
Seri Begawan

Jakarta

Mariana Trench

Equator

INDONESIA

International Date Line

154° W

Dili

Melekeok

SOLOMON
ISLANDS

Southern

Port Moresby

Honiara

Pacific Ocean

AUSTRALIA

VANUATU

Port Vila

Suva

FIJI

Canberra

Wellington

MOBILE MARITIME LAUNCH PLATFORM

Name
Sea Launch S.A.

Home Base
Slavyanka, Russia

Owner/Operator
S7 Group

Elevation
0 m

Coordinates
154°W

Time zone
GMT+10

Launches
36 satellites

Completion
1995

Sea Launch used the same idea as San Marco, but was much larger. It used the powerful Zenit rocket, built in Dnepropetrovsk, Ukraine, with Russian RD-171 rocket engines. Atop Zenit was the Energiya design bureau's blok D upper stage, able to send commercial communications satellites of 5.25 tonnes into geosynchronous orbit. Called Zenit 3SL (three stages, Sea Launch), it was a post-Cold War Russian-American-Ukrainian project intended as a money-maker for America's Boeing (40 per cent), Russia's Energiya (25 per cent), Norway's Kværner (20 per cent), and Ukraine's Yuzhnoye (15 per cent) companies.

There were two elements. First was the 34,000-tonne rocket transport, fuel, and assembly ship, Sea Commander, built in Govan, Scotland. It was big, measuring 203 m in length and 32 m in width, and able to fit three Zenits in a hangar 70 m long and 12 m deep. In 1998, Sea Commander travelled to its home port of Long Beach, California, where Boeing cleared an old naval depot. Sea Commander's role was to load the Zenits, bring them 1,600 km across the Pacific south of Hawaii near Kiritimati, transfer them to the launch platform, and move 5 km away for launch control and tracking. The second element was the launch platform, Odyssey – a decommissioned self-propelled, semi-submersible Norwegian oil rig, 131 m long, 78 m wide, and 70 m tall with sunken pontoons. It was refitted in Stavanger and Vyborg with stronger legs, new power systems, plumbing for rocket fuel, a helipad, a flame trench, crew quarters, and a crane. Odyssey barely made it under the new Øresund Bridge on its 119-day journey to the Pacific. Rockets were shipped from Dnepropetrovsk and fuel from Russia. For launch, the hangar roof rolled back, Zenit was raised

The main control ship docked alongside
the semi-submersible launch platform.
Source: Sea Launch

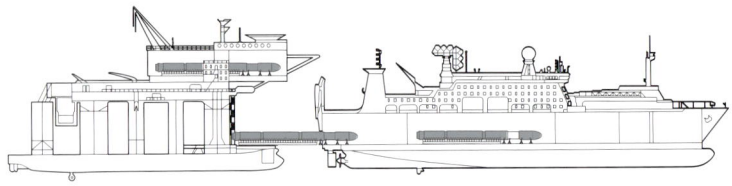

Scaled diagram of Sea Launch rocket systems
Source: Lauren Carmona

Sea Launch Odyssey   **Sea Launch Commander**   **China: De Bo 3**

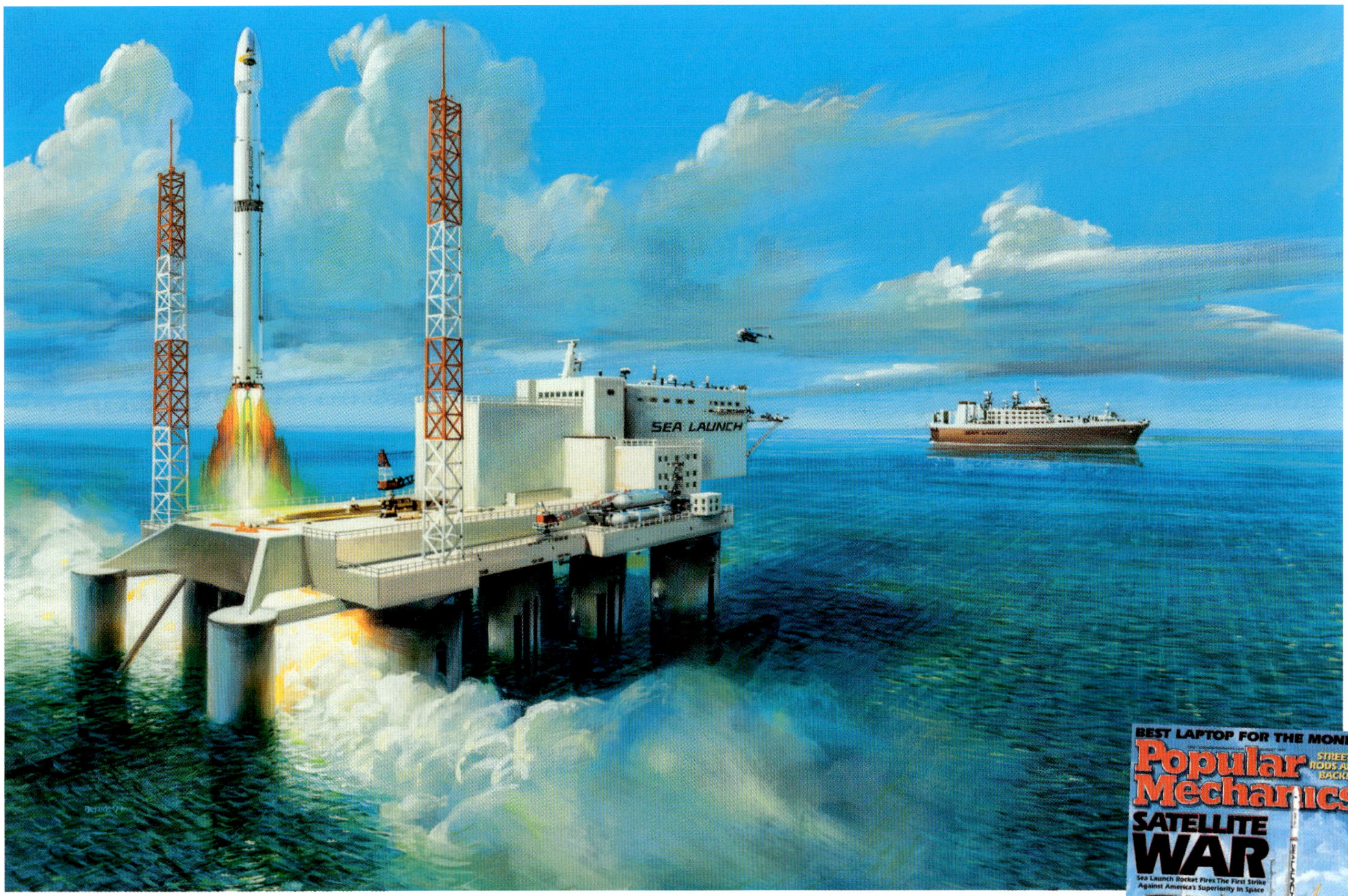
SEA LAUNCH

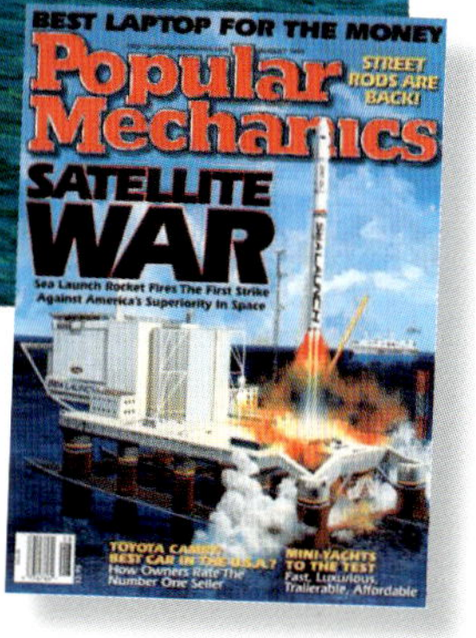
BEST LAPTOP FOR THE MONEY
Popular Mechanics
STREET RODS ARE BACK!
SATELLITE WAR

Cover illustration by John Berkey, which appeared in the magazine *Popular Mechanics*.
Source: Popular Mechanics 8/1999

hydraulically to the vertical, Odyssey was evacuated, and the launch was entirely automated and monitored from Sea Commander. Sea Launch made a perfect demonstration mission in March 1999, with its first operational mission coming that October. By 2014, Sea Launch had made 36 launches for American, European, and Asian customers (with three failures). The project then halted when Zenit production stopped. Sea Launch was bought for \$150m by the Russian S7 Airlines in April 2018. After lengthy negotiations for their departure from Long Beach, both vessels crossed the Pacific in 2020 to reach Slavyanka, Primorsky, at the southernmost tip of the Russian far east, for outfitting with the new Soyuz 7 rocket. In poor, stripped out condition, restoration would take some time.

Next double page:
An aerial view of Hong Kong's Xin Guang Hua heavy load carrier carrying the Odyssey mobile maritime spacecraft launch platform of the Sea Launch international spacecraft launch service that arrived from the United States at Slavyanka Port.
Source: Picture Alliance / Dmitry Yefremov / TASS

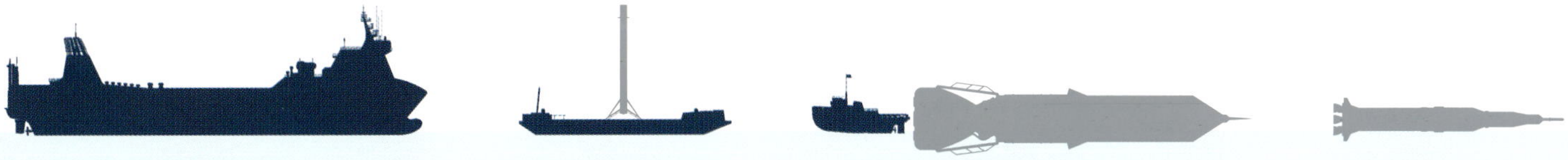

Blue Origin: Jacklyn
SpaceX: Drone ship
NASA: Sea Dragon (concept)
NASA: Saturn V

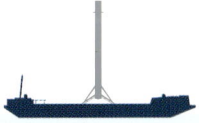

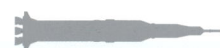

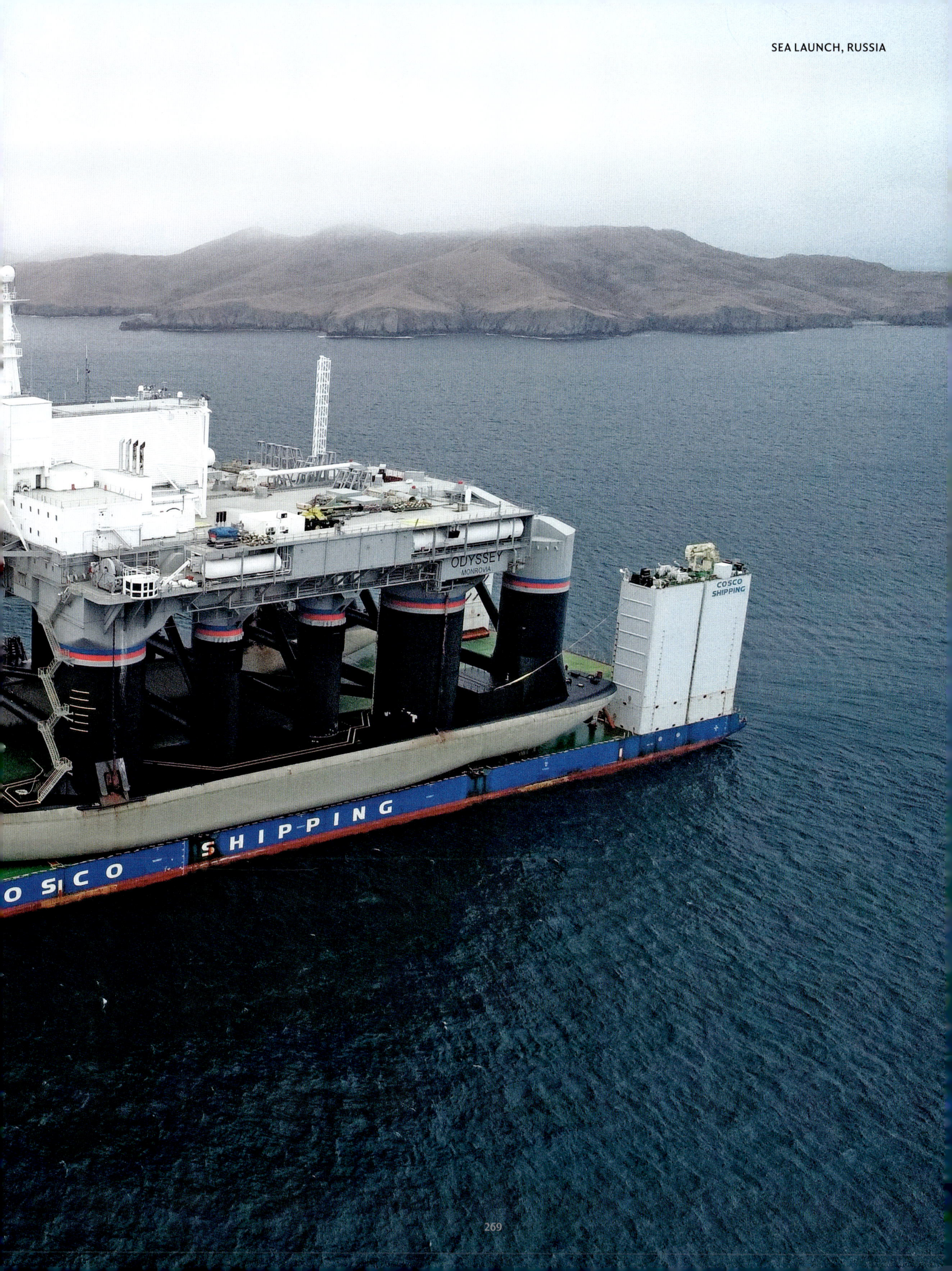
ODYSSEY
MONROVIA
COSCO
SHIPPING
COSCO
SHIPPING

INDEX

BIOGRAPHIES

Author

Brian Harvey is a writer and broadcaster on spaceflight who lives in Dublin, Ireland. He has a degree in history and political science from the University of Dublin (Trinity College) and an MA from University College Dublin. His first book was *Race into Space – The Soviet Space Programme* (Ellis Horwood, 1988), followed by further publications on the Russian, Chinese, European, Indian, and Japanese space programmes. His books and book chapters have been translated into Russian, Chinese, and Korean. He has recently completed *China in Space – The Great Leap Forward* (2nd, edition, Praxis-Springers, 2019), now being translated into Chinese for publication there, and *European-Russian Space Cooperation – From De Gaulle to ExoMars* (Praxis-Springer, 2021). He has broadcast on BBC, Canadian Broadcasting Corporation (CBC), Voice of America, and China Television (*Dialogue - ideas matter*). He has contributed to films by CBC (Mir), Danish television (closed ecological systems), and Australian television (the H-II Japanese rocket), subsequently shown on the Discovery channel. He has been interviewed for *The Observer, Christian Science Monitor, The Guardian, Наша Газета (Nasha Gazeta) Le Scienze, Ça M'intéresse,* and *The Hindu*. He has written articles on spaceflight from the 1970s in magazines such as *Orbit, Astronomy and Space, Go Taikonauts!,* and *Spaceflight*, and for newspapers such as the *Sunday Press* and the *Irish Independent*. His articles have been published in *Astronomy Now, Space Quarterly, Space Policy, Journal of the British Interplanetary Society, Zenit, Quest,* and by the Leibniz Society in Berlin. He is a Fellow of the British Interplanetary Society (FBIS) and co-chaired its annual Sino-Russian forum for a number of years. He contributed to the review of space policy by the Department of Business and Skills that led to the establishment of the UK Space Agency (UKSA). Marking the 50th anniversary of manned spaceflight, he opened the UK Yuri Gagarin exhibition in Edinburgh, organised by the Princess Dashkova Centre of the University of Edinburgh.

Co-Author

Gurbir Singh is a UK-based non-fiction writer specialising in space. He studied science and computing and holds a science and an arts degree. Once keen on aviation, he has a private pilot's licence for the UK, USA, and Australia. He was one of 13,000 unsuccessful applicants responding to the 1989 advert *'Astronaut wanted. No experience necessary'* to become the first British astronaut. Helen Sharman was eventually selected and flew on the Soviet space station Mir in 1991. He has written articles for *The Space Review, Go Taikonauts!, Journal of the British Interplanetary Society,* and *Spaceflight* and has been interviewed for BBC Manchester, Deutsche Welle, and the BBC World Service. In late 2018, he stopped working full time as a cyber security consultant to spend more time on his writing. He is also the publisher of the website www.astrotalkuk.org, a not-for-profit astronomy podcast established in 2008. In 2011, he published his first book, *Yuri Gagarin in London and Manchester*. The book traces the visit of the world's first spaceman's five days in England, with firsthand accounts from the people who saw and met him. His second book, *The Indian Space Programme,* published in October 2017, is a detailed account of the origin of India's space programme, its achievements, and future ambitions. His third book, *India's Forgotten Rocket Pioneer,* is a biographic account of the life and work of Stephen H. Smith, who experimented with rockets between the 1930s and 1940 in Calcutta, India.

Editor

Paul Meuser is a researcher and artist dealing with both the interstellar ambitions and earthly artifacts of humanities leap into space. Born in 1996 in Berlin, Germany, he holds a BFA with a concentration in 'computation, technology, and culture' from the Rhode Island School of Design, and an MArch from the Yale School of Architecture. In 2019, he published the *Architectural Guide Moon*, exploring the first permanent human footprints on our grey neighbour (German edition: 2019/DOM publishers; Lithuanian edition: 2022/LAPAS publishing house; Russian edition: in progress). Together with the Yale Centre for Collaborative Arts and Media and the Massachusetts Institute of Technology (MIT), he took part in an experimental ZeroG flight with scientists and artists. Experiencing the absence of gravity on oneself was truly eye-opening, connecting research and the experience itself. His earthly work focuses on robotics and its applications beyond their conventional uses. His most recent explorations dealt with the use of soft robotics to introduce age and mortality into machines and satellites. He currently works as an architect and design engineer in Tunisia.

Maps

Katrin Soschinski has been working as a freelance cartographer since 2007, both on a freelance and project basis for publishers, agencies, and planning offices. One focus of her work is textbook cartography, but cartographic support for public sector spatial planning is also part of her spectrum. She concentrates mainly on the print sector. After studying geography and cartography in Bochum and completing a cartographic publishing traineeship at the Bertelsmann subsidiary *wissenmedia* in Stuttgart, she lived and worked in Berlin for 10 years. At the end of 2019, she moved to the western Ruhr region, where the focus of her life has been ever since. Her cartography is influenced by her interest in design and art, while at the same time shaped by the hope for a free, safe, and healthy world in the future.

The *Deutsche Nationalbibliothek* lists this publication in the *Deutsche Nationalbibliografie*; detailed bibliographic data are available at http://dnb.d-nb.de

ISBN 978-3-86922-758-0

Picture editing
Paul Meuser

Proofreading
Sandie Kestell

Design
Masako Tomokiyo

Pre-press
Tiger Printing (Hong Kong), Co. Ltd
www.tigerprinting.hk

Printing
Bilnet Matbaacılık ve Yayıncılık A. Ş., Istanbul
www.bilnet.net.tr

DOM
publishers

1975

Tanegashima
Japan

1979

Sriharikota
India

1984

Xichang
China

1988

Taiyuan
China

1988

Palmachim
Israel

1997

Alcântara
Brazil

2001

Kodiak
United States